The **ESSENTIALS**®

Math Made Nice-n-Easy III

Factoring • Ratios
Linear Equations
Proportions • Variations
Functions

> "MATH MADE NICE-n-EASY III" is one in a series of books designed to make the learning of math interesting and fun. For help with additional math topics, see the complete series of "MATH MADE NICE-n-EASY" titles.

Research & Education Association
61 Ethel Road West
Piscataway, New Jersey 08854

THE ESSENTIALS®
MATH MADE NICE-N-EASY III

Printed in the United States of America

Library of Congress Catalog Card Number 99-70140

International Standard Book Number 0-87891-202-9

ESSENTIALS is a registered trademark of
Research & Education Association, Piscataway, New Jersey 08854

WHAT "MATH MADE NICE-N-EASY" WILL DO FOR YOU

The "Math Made Nice-n-Easy" series simplifies the learning and use of math and lets you see that math is actually interesting and fun. This series of books is for people who have found math scary, but who nevertheless need some understanding of math without having to deal with the complexities found in most math textbooks.

The "Math Made Nice-n-Easy" series of books is useful for students and everyone who needs to acquire a basic understanding of one or more math topics. For this purpose, the series is divided into a number of books which deal with math in an easy-to-follow sequence beginning with basic arithmetic, and extending through pre-algebra, algebra, and calculus. Each topic is described in a way that makes learning and understanding easy.

Almost everyone needs to know at least some math at work, or in a course of study.

For example, almost all college entrance tests and professional exams require solving math problems. Also, almost all occupations (waiters, sales clerks, office people) and all crafts (carpentry, plumbing, electrical) require some ability in math problem solving.

The "Math Made Nice-n-Easy" series helps the reader grasp quickly the fundamentals that are needed in using math. The reader is led by the hand, step-by-step, through the various concepts and how they are used.

By acquiring the ability to use math, the reader is encouraged to further his/her skills and to forget about any initial math fears.

The "Math Made Nice-n-Easy" series includes material originated by U.S. Government research and educational efforts. The research was aimed at devising tutoring and teaching methods for educating government personnel lacking a technical and/or mathematical background. Thanks for these efforts are due to the U.S. Bureau of Naval Personnel Training.

Dr. Max Fogiel
Program Director

Contents

Chapter 13

RATIOS, PROPORTIONS, AND VARIATIONS 375

Chapter 14

DEPENDENCE, FUNCTIONS AND FORMULAS 403

CHAPTER 10

FACTORING POLYNOMIALS

A factor of a quantity N, as defined in chapter 2 of this course, is any expression which can be divided into N without producing a remainder. Thus 2 and 3 are factors of 6, and the factors of 5x are 5 and x. Conversely, when all of the factors of N are multiplied together, the product is N. This definition is extended to include polynomials.

The factors of a polynomial are two or more expressions which, when multiplied together, give the polynomial as a product. For example, 3, x, and $x^2 - 4$ are factors of $3x^3 - 12x$, as the following equation shows:

$$(3)(x)(x^2 - 4) = 3x^3 - 12x$$

The factors 3 and x, which are common to both terms of the polynomial $3x^3 - 12x$, are called COMMON FACTORS.

The distributive principle, mentioned in chapters 3 and 9 of this course, is an important part of the concept of factoring. It may be stated as follows:

If the sum of two or more quantities is multiplied by a third quantity, the product is found by applying the multiplier to each of the original quantities separately and summing the resulting expressions. It is this principle which allows us to separate common factors from the terms of a polynomial.

Just as with numbers, an algebraic expression is a prime factor if it has no other factors except itself and 1. The factor $x^2 - 4$ is not prime, since it can be separated into $x - 2$ and $x + 2$. The factors $x - 2$ and $x + 2$ are both prime factors, since they cannot be separated into other factors.

The process of finding the factors of a polynomial is called FACTORING. An expression is said to be factored completely when it has been separated into its prime factors. The polynomial $3x^3 - 12x$ is factored completely as follows:

$$3x^3 - 12x = 3x(x - 2)(x + 2)$$

COMMON FACTORS

Factoring any polynomial begins with the removal of common factors. Notice that "removal" of a factor does not mean discarding it. To remove a factor is to insert parentheses and move the factor outside the parentheses as a common multiplier. The removal of common factors proceeds as follows:

1. Inspect the polynomial and find the factors which are common to all terms. These common factors, multiplied together, comprise the "largest common factor."

2. Mentally divide each term of the polynomial by the largest common factor and write the quotients within a set of parentheses.

3. Write the largest common factor outside the parentheses as a common multiplier.

For example, the expression $x^2y - xy^2$ contains xy as a factor of each term. Therefore, it is factored as follows:

$$x^2y - xy^2 = xy(x - y)$$

Other examples of factoring by the removal of common factors are found in the following expressions:

$$6m^4n + 3m^3n^2 - 3m^2n^3 = 3m^2n(2m^2 + mn - n^2)$$

$$-5z^2 - 15z = -5z(z + 3)$$

$$7x - 7y + 7z = 7(x - y + z)$$

In selecting common factors, always remove as many factors as possible from each term in order to factor completely. For example, x is a factor of $3ax^2 - 3ax$, so that $3ax^2 - 3ax$ is equal to x(3ax -3a). However, 3 and a are also factors. Thus the largest common factor is 3ax. When factored completely, the expression is as follows:

$$3ax^2 - 3ax = 3ax(x - 1)$$

Practice problems: Remove the common factors:

1. $y^2 - y$

2. $a^3b^2 - a^2b^2$

3. $2b^3 - 8b^2 - 6b$

4. $6mn^2 + 30m^2n$

5. $\frac{2}{3}x - \frac{1}{3}y + \frac{1}{3}$

Answers:

1. $y(y - 1)$

2. $a^2b^2(a - 1)$

3. $2b(b^2 - 4b - 3)$

4. $6mn(n + 5m)$

5. $\frac{1}{3}(2x - y + 1)$

LITERAL EXPONENTS

It is frequently necessary to remove common factors involving literal exponents; that is, exponents composed of letters rather than numbers. A typical expression involving literal exponents is $x^{2a} + x^a$, in which x^a is a common factor. The factored form is $x^a(x^a + 1)$. Another example of this type is $a^{m+n} + 2a^m$. Remember that a^{m+n} is equivalent to $a^m \cdot a^n$. Thus the factored form is as follows:

$$a^{m+n} + 2a^m = a^m \cdot a^n + 2a^m$$
$$= a^m(a^n + 2)$$

BINOMIAL FORM

The distinctions between monomial, binomial, and trinomial factors are discussed in detail in chapter 9 of this course. An expression such as $a(x + y) + b(x + y)$ has a common factor in binomial form. The factor $(x + y)$ can be removed from both terms, with the following result:

291

$$a(x + y) + b(x + y) = (x + y)(a + b)$$

Sometimes it is easier to see this if a single letter is substituted temporarily for the binomial. Thus, let $(x + y) = n$, so that $a(x + y) + b(x + y)$ reduces to $(an + bn)$. The factored form is $n(a + b)$, which becomes $(x + y)(a + b)$ when n is replaced by its equal, $(x + y)$.

Another form of this type is $x(y - z) - w(z - y)$. Notice that this expression could be factored easily if the binomial in the second term were $(y - z)$. We can show that $-w(z - y)$ is equivalent to $+w(y - z)$, as follows:

$$-w(z - y) = -w\ [(-1) \cdot (-1) \cdot z + (-1) \cdot y]$$
$$= -w\ \{(-1)\ [(-1)\ z + y]\}$$
$$= (-w)(-1)\ [-z + y]$$
$$= +w(y - z)$$

Substituting $+w(y - z)$ for $-w(z - y)$ in the original expression, we may now factor as follows:

$$x(y - z) -w(z - y) = x(y - z) + w(y - z)$$
$$= (y - z)(x + w)$$

In factoring an expression such as $ax + bx + ay + by$, common monomial factors are removed first, as follows:

$$ax + bx + ay + by = x(a + b) + y(a + b)$$

Having removed the common monomial factors, we then remove the common binomial factor to obtain $(a + b)(x + y)$.

Notice that we could have rewritten the expression as $ax + ay + bx + by$, based on the commutative law of addition, which states that

the sum of two or more terms is the same regardless of the order in which they are arranged. The first step in factoring would then
produce $a(x + y) + b(x + y)$ and the final form
would be $(x + y)(a + b)$. This is equivalent to
$(a + b)(x + y)$, by the commutative law of multiplication, which states that the product of two
or more factors is the same regardless of the
order in which they are arranged.

Practice problems. Factor each of the following:

1. $x^{3a} + 3x^{2a}$

2. $xy^2 + y + x^2y + x$

3. $e^x + 4e^{4x}$

4. $7(x^2 + y^2) - 3z(x^2 + y^2)$

5. $a^2 + ab - ac - cb$

6. $\frac{1}{2}e^2r - \frac{1}{6}er^2$

7. $a^{x+2} + a^2$

8. $xy - 3x - 2y + 6$

Answers:

1. $x^{2a}(x^a + 3)$

2. $(xy + 1)(x + y)$

3. $e^x(1 + 4e^{3x})$

4. $(x^2 + y^2)(7 - 3z)$

5. $(a + b)(a - c)$

6. $\frac{1}{2}er(e - \frac{1}{3}r)$

7. $a^2(a^x + 1)$

8. $(y - 3)(x - 2)$

BINOMIAL FACTORS

After any common factor has been removed from a polynomial, the remaining polynomial factor must be examined further for other factors. Skill in factoring is principally the ability to recognize certain types of products such as the square of a sum or difference. Therefore, it is important to be familiar with the special products discussed in chapter 9.

DIFFERENCE OF TWO SQUARES

In chapter 9 we learned that the product of the sum and difference of two numbers is the difference of their squares. Thus, $(a + b)(a - b) = a^2 - b^2$. Conversely, if a binomial is the difference of two squares, its factors are the sum and difference of the square roots. For example, in $9a^2 - 4b^2$ both $9a^2$ and $4b^2$ are perfect squares. The square roots are 3a and 2b, respectively. Connect these square roots with a plus sign to get one factor of $9a^2 - 4b^2$ and with a minus sign to get the other factor. The two binomial factors are 3a - 2b and 3a + 2b. Therefore, factored completely, the binomial can be written as follows:

$$9a^2 - 4b^2 = (3a - 2b)(3a + 2b)$$

We may check to see if these factors are correct by multiplying them together to see if their product is the original binomial.

The expression $20x^3y - 5xy^3$ reduces to the difference of two squares after the common factor $5xy$ is removed. Completely factored, this expression produces the following:

$$20x^3y - 5xy^3 = 5xy(4x^2 - y^2)$$
$$= 5xy(2x - y)(2x + y)$$

Other examples that show the difference of two squares in factored form are as follows:

$$49 - 16 = (7 + 4)(7 - 4)$$
$$16a^2 - 4x^2 = 4(4a^2 - x^2)$$
$$= 4(2a + x)(2a - x)$$
$$4x^2y - 9y = y(4x^2 - 9)$$
$$= y(2x + 3)(2x - 3)$$

Practice problems: Factor each of the following:

1. $a^2 - b^2$ 5. $x^2 - y^2$

2. $b^2 - 9$ 6. $y^2 - 36$

3. $a^2b^2 - 1$ 7. $1 - 4y^2$

4. $a^2 - 144$ 8. $9a^2 - 16$

Answers:

1. $(a + b)(a - b)$ 5. $(x + y)(x - y)$

2. $(b + 3)(b - 3)$ 6. $(y + 6)(y - 6)$

3. $(ab + 1)(ab - 1)$ 7. $(1 + 2y)(1 - 2y)$

4. $(a + 12)(a - 12)$ 8. $(3a + 4)(3a - 4)$

SPECIAL BINOMIAL FORMS

Special cases involving binomial expressions are frequently encountered. All such expressions may be factored by reference to general formulas, but these formulas are beyond the scope of this course. For our purposes, analysis of some special cases will be sufficient.

Even Exponents

When the exponents on both terms of the binomial are even, the expression may be treated as the sum or difference of two squares. For example, $x^6 - y^6$ can be rewritten as $(x^3)^2 - (y^3)^2$ which results in the following factored form:

$$x^6 - y^6 = (x^3 - y^3)(x^3 + y^3)$$

In general, a binomial with even exponents has the form $x^{2m} \pm y^{2n}$, since all even numbers have 2 as a factor. If the connecting sign is positive, the expression may not be factorable; for example, $x^2 + y^2$, $x^4 + y^4$, and $x^8 + y^8$ are all nonfactorable binomials. If the connecting sign is negative, a binomial with even exponents is factorable as follows:

$$x^{2m} - y^{2n} = (x^m - y^n)(x^m + y^n)$$

A special case which is particularly important because it occurs so often is the binomial which has the numeral 1 as one of its terms. For example, $x^4 - 1$ is factorable as the difference of two squares, as follows:

$$x^4 - 1 = (x^2 - 1)(x^2 + 1)$$
$$= (x - 1)(x + 1)(x^2 + 1)$$

Odd Exponents

Two special cases involving odd exponents are of particular importance. These are the sum of two cubes and the difference of two cubes. Examples of the sum and difference of two cubes, showing their factored forms, are as follows:

$$x^3 + y^3 = (x + y)(x^2 - xy + y^2)$$
$$x^3 - y^3 = (x - y)(x^2 + xy + y^2)$$

Notice that each of these factored forms involves a first degree binomial factor $((x + y)$ in the first case and $(x - y)$ in the second). The connecting sign in the first degree binomial factor corresponds to the connecting sign in the original unfactored binomial.

We are now in a position to give the completely factored form of $x^6 - y^6$, as follows:

$$x^6 - y^6 = (x^3 - y^3)(x^3 + y^3)$$
$$= (x - y)(x^2 + xy + y^2)$$
$$(x + y)(x^2 - xy + y^2)$$

In general, $(x + y)$ is a factor of $(x^n + y^n)$ if n is odd. If n is even, $(x^n + y^n)$ is not factorable unless it can be expressed as the sum of two cubes. When the connecting sign is negative, the binomial is always factorable if n is a whole number greater than 1. That is, $(x - y)$

is a factor of $(x^n - y^n)$ for both odd and even values of n.

The special case in which one of the terms of the binomial is the numeral 1 occurs frequently. An example of this is $x^3 + 1$, which is factorable as the sum of two cubes, as follows:

$$x^3 + 1 = (x + 1)(x^2 - x + 1)$$

In a similar manner, $1 + x^6$ can be treated as the sum of two cubes and factored as follows:

$$1 + x^6 = 1 + (x^2)^3$$
$$= (1 + x^2)(1 - x^2 + x^4)$$

Practice problems. In each of the following problems, factor completely:

1. $x^4 - y^4$ 4. $x^3 - y^3$ 7. $1 - x^4$

2. $m^3 + n^3$ 5. $a^9 - b^9$ 8. $x^6 + 1$

3. $x^6 - y^6$ 6. $x^{2a} - y^{2b}$ 9. $1 - x^3$

Answers:

1. $(x + y)(x - y)(x^2 + y^2)$
2. $(m + n)(m^2 - mn + n^2)$
3. $(x + y)(x - y)(x^2 + xy + y^2)(x^2 - xy + y^2)$
4. $(x - y)(x^2 + xy + y^2)$
5. $(a - b)(a^2 + ab + b^2)(a^6 + a^3b^3 + b^6)$
6. $(x^a - y^b)(x^a + y^b)$
7. $(1 + x^2)(1 - x)(1 + x)$

8. $(x^2 + 1)(x^4 - x^2 + 1)$

9. $(1 - x)(1 + x + x^2)$

TRINOMIAL SQUARES

A trinomial that is the square of a binomial is called a TRINOMIAL SQUARE. Trinomials that are perfect squares factor into either the square of a sum or the square of a difference. Recalling that $(x + y)^2 = x^2 + 2xy + y^2$ and $(x - y)^2 = x^2 - 2xy + y^2$, the form of a trinomial square is apparent. The first term and the last term are perfect squares and their signs are positive. The middle term is twice the product of the square roots of these two numbers. The sign of the middle term is plus if a sum has been squared; it is minus if a difference has been squared.

The polynomial $16x^2 - 8xy + y^2$ is a trinomial in which the first term, $16x^2$, and the last term, y^2, are perfect squares with positive signs. The square roots are $4x$ and y. Twice the product of these square roots is $2(4x)(y) = 8xy$. The middle term is preceded by a minus sign indicating that a difference has been squared. In factored form this trinomial is as follows:

$$16x^2 - 8xy + y^2 = (4x - y)^2$$

To factor the trinomial, we simply take the square roots of the end terms and join them with a plus sign if the middle term is preceded by a plus or with a minus if the middle term is preceded by a minus.

The terms of a trinomial may appear in any order. Thus, $8xy + y^2 + 16x^2$ is a trinomial square and may be factored as follows:

$$8xy + y^2 + 16x^2 = 16x^2 + 8xy + y^2 = (4x + y)^2$$

Practice problems. Among the following expressions, factor those which are trinomial squares:

1. $y^2 - 8y + 16$

2. $16y^2 + 30x + 9$

3. $36 + 12x + x^2$

4. $a^2 + 2ab + b^2$

5. $12y + 9y^2 - 4$

6. $4x^2 + y^2 + 4xy$

7. $9 - 6cd + c^2d^2$

8. $x^4 + 4x^2 + 4$

Answers:

1. $(y - 4)^2$

2. Not a trinomial square

3. $(6 + x)^2$

4. $(a + b)^2$

5. Not a trinomial square

6. $(2x + y)^2$

7. $(3 - cd)^2$

8. $(x^2 + 2)^2$

SUPPLYING THE MISSING TERM

Skill in recognizing trinomial squares may be improved by practicing the solution of problems which require supplying a missing term. For example, the expression $y^2 + (?) + 16$ can be made to form a perfect trinomial square by supplying the correct term to fill the parentheses.

The middle term must be twice the product of the square roots of the two perfect square terms; that is, $(2)(4)(y)$, or $8y$. Check: $y^2 + 8y + 16 = (y + 4)^2$. The missing term is $8y$.

Suppose that we wish to supply the missing term in $16x^2 + 24xy + (?)$ so that the three terms will form a perfect trinomial square. The square root of the first term is $4x$. One-half the middle term is $12xy$. Divide $12xy$ by $4x$. The result is $3y$ which is the square root of the last term. Thus, our missing term is $9y^2$. Checking, we find that $(4x + 3y)^2 = 16x^2 + 24xy + 9y^2$.

Practice problems. In each of the following problems, supply the missing term to form a perfect trinomial square:

1. $x^2 + (?) + y^2$

2. $t^2 + (?) + 25$

3. $9a^2 - (?) + 25b^2$

4. $4m^2 + 16m + (?)$

5. $x^2 + 4x + (?)$

6. $c^2 - 6cd + (?)$

Answers:

1. $2xy$

2. $10t$

3. $30ab$

4. 16

5. 4

6. $9d^2$

OTHER TRINOMIALS

It is sometimes possible to factor trinomials that are not perfect squares. Following are some examples of such trinomials, and the expressions of which they are products:

1. $(x + 3)(x + 4) = x^2 + 7x + 12$

2. $(x - 3)(x - 4) = x^2 - 7x + 12$

3. $(x - 3)(x + 4) = x^2 + x - 12$

4. $(x + 3)(x - 4) = x^2 - x - 12$

It is apparent that trinomials like these may be factored into binomials as shown. Notice how the trinomial in each of the preceding examples is formed. The first term is the square of the common term of the binomial factors. The second term is the algebraic sum of their unlike terms times their common term. The third term is the product of their unlike terms.

Such trinomials may be factored as the product of two binomials if there are two numbers such that their algebraic sum is the coefficient of the middle term and their product is the last term.

For example, let us factor the expression $x^2 - 12x + 32$. If the expression is factorable, there will be a common term, x, in each of the binomial factors. We begin factoring by placing this term within each set of parentheses, as follows:

$$(x \quad)(x \quad)$$

Next, we must find the other terms that are to go in the parentheses. They will be two numbers such that their algebraic sum is -12 and their product is +32. We see that -8 and -4 satisfy the conditions. Thus, the following expression results:

$$x^2 - 12x + 32 = (x - 8)(x - 4)$$

It is of value in factoring to note some useful facts about trinomials. If both the second and third terms of the trinomial are positive, the signs of the terms to be found are positive as in example 1 of this section. If the second term is negative and the last is positive, both terms to be found will be negative as in example 2. If the third term of the trinomial is negative, one of the terms to be found is positive and the other is negative as in examples 3 and 4. Concerning this last case, if the second term is positive as in example 3, the positive term in the factors has the greater numerical value. If the second term is negative as in example 4, the negative term in the factors has the greater numerical value.

It should be remembered that not all trinomials are factorable. For example, $x^2 + 4x + 2$ cannot be factored since there are no two rational numbers whose product is 2 and whose sum is 4.

Practice problems. Factor completely, in the following problems:

1. $y^2 + 15y + 50$

2. $y^2 - 2y - 24$

3. $x^2 + 8x - 48$

4. $x^2 - 4x - 60$

5. $x^2 - 12x - 45$

6. $x^2 - 15x + 56$

7. $x^2 + 2x - 48$

8. $x^2 + 14x + 24$

Answers:

1. $(y + 5)(y + 10)$

2. $(y - 6)(y + 4)$

3. $(x + 12)(x - 4)$

4. $(x - 10)(x + 6)$

5. $(x - 15)(x + 3)$

6. $(x - 7)(x - 8)$

7. $(x - 6)(x + 8)$

8. $(x + 12)(x + 2)$

Thus far we have considered only those expressions in which the coefficient of the first term is 1. When the coefficient of the first term is other than 1, the expression can be factored as shown in the following example:

$$6x^2 - x - 2 = (2x + 1)(3x - 2)$$

Although this result can be obtained by the trial and error method, the following procedure saves time and effort. First, find two numbers whose sum is the coefficient of the second term (-1 in this example) and whose product is equal to the product of the third term and the coefficient of the first term (in this example, $(6)(-2)$ or -12). By inspection, the desired numbers are found to be -4 and +3. Using these two numbers as coefficients for x, we can rewrite the original expression as $6x^2 - 4x + 3x - 2$ and factor as follows:

$$6x^2 - 4x + 3x - 2 = 2x(3x - 2) + 1(3x - 2)$$
$$= (2x + 1)(3x - 2)$$

Practice problems. Factor completely, in the following problems:

1. $2x^2 + 13x + 21$ 3. $15x^2 - 16x - 7$

2. $16x^2 + 26x + 3$ 4. $12x^2 - 8x - 15$

Answers:

1. $(2x + 7)(x + 3)$ 3. $(3x + 1)(5x - 7)$

2. $(2x + 3)(8x + 1)$ 4. $(6x + 5)(2x - 3)$

REDUCING FRACTIONS TO LOWEST TERMS

There are many useful applications of factoring. One of the most important is that of simplifying algebraic fractions. Fractions that contain algebraic expressions in the numerator or denominator, or both, can be reduced to lower terms, if there are factors common to numerator and denominator. If the terms of a fraction are monomials, common factors are immediately apparent, as in the following expression:

$$\frac{3x^2y}{6xy} = \frac{3xy(x)}{3xy(2)} = \frac{x}{2}$$

If the terms of a fraction are polynomials, the polynomials must be factored in order to recognize the existence of common factors, as in the following two examples:

1. $\dfrac{a - b}{a^2 - 2ab + b^2} = \dfrac{a - b}{(a - b)(a - b)} = \dfrac{1}{(a - b)}$

2. $\dfrac{4x^2 - 9}{6x^2 - 9x} = \dfrac{(2x + 3)(2x - 3)}{3x(2x - 3)} = \dfrac{(2x + 3)}{3x}$

Notice that without the valuable process of factoring, we would be forced to use the fractions in their more complicated form. When there are factors common to both numerator and denominator, it is obviously more practical to cancel them (first using the factoring process) before proceeding.

Practice problems. Reduce to lowest terms in each of the following:

1. $\dfrac{12}{6x + 12}$

4. $\dfrac{y^2 - 25}{y^2 - 8y + 15}$

2. $\dfrac{a^2 - b^2}{a^2 - 2ab + b^2}$

5. $\dfrac{a^2 - 5a - 24}{a^2 - 64}$

3. $\dfrac{y^2 - 14y + 45}{y^2 - 8y - 9}$

6. $\dfrac{4x^2y - 9y}{4x^2 + 12x + 9}$

Answers:

1. $\dfrac{2}{x + 2}$

4. $\dfrac{y + 5}{y - 3}$

2. $\dfrac{a + b}{a - b}$

5. $\dfrac{a + 3}{a + 8}$

3. $\dfrac{y - 5}{y + 1}$

6. $\dfrac{y(2x - 3)}{2x + 3}$

OPERATIONS INVOLVING FRACTIONS

Addition, subtraction, multiplication, and division operations involving algebraic fractions are often simplified by means of factoring, whereas they would be quite complicated without the use of factoring.

MULTIPLYING FRACTIONS

Multiplication of fractions that contain polynomials is similar to multiplication of fractions that contain only arithmetic numbers. If this fact is kept in mind, the student will have little difficulty in mastering multiplication in algebra. For instance, we recall that to multiply a fraction by a whole number, we simply multiply the numerator by the whole number. This is illustrated in the following example:

Arithmetic: $4 \times \dfrac{3}{17} = \dfrac{12}{17}$

Algebra: $(x - 4) \cdot \dfrac{3}{x^2 - 5} = \dfrac{3x - 12}{x^2 - 5}$

Sometimes the work may be simplified by factoring and canceling before carrying out the multiplication. The following example illustrates this:

$$(2a - 8) \cdot \frac{3}{a^2 - 8a + 16} = \frac{2\cancel{(a - 4)}}{1} \cdot \frac{3}{(a - 4)\cancel{(a - 4)}}$$

$$= \frac{2(3)}{a - 4} = \frac{6}{a - 4}$$

When the multiplier is a fraction, the rules of arithmetic remain applicable—that is, multiply numerators together and denominators together. This is illustrated as follows:

Arithmetic: $\dfrac{4}{5} \times \dfrac{2}{3} = \dfrac{8}{15}$

Algebra: $\dfrac{a + b}{a - b} \cdot \dfrac{a}{a - b} = \dfrac{a(a + b)}{(a - b)^2}$

Where possible, the work may be considerably reduced by factoring, canceling, and then carrying out the multiplication, as in the following example:

307

$$\frac{x^2 - 2x + 1}{x^2 - 9} \cdot \frac{x^2 + x - 6}{x^2 - 1}$$

$$= \frac{(x - 1)\cancel{(x - 1)}}{\cancel{(x + 3)}(x - 3)} \cdot \frac{\cancel{(x + 3)}(x - 2)}{(x + 1)\cancel{(x - 1)}}$$

$$= \frac{(x - 1)(x - 2)}{(x - 3)(x + 1)} = \frac{x^2 - 3x + 2}{x^2 - 2x - 3}$$

Although the factors may be multiplied to form two trinomials as shown, it is usually sufficient to leave the answer in factored form.

Practice problems. In the following problems, multiply as indicated:

1. $5a^2 \cdot \dfrac{3b}{a + b}$

2. $\dfrac{x + y}{x^2} \cdot \dfrac{x - y}{x - 1}$

3. $\dfrac{a^2 + 2ab + b^2}{a^2 - b^2} \cdot \dfrac{6a}{3a + 3b}$

4. $\dfrac{a - 1}{2a^2 + 4a + 2} \cdot \dfrac{(a + 1)^2}{a - 1}$

Answers:

1. $\dfrac{15a^2b}{a + b}$

2. $\dfrac{x^2 - y^2}{x^3 - x^2}$

3. $\dfrac{6a}{3(a - b)}$

4. $\dfrac{1}{2}$

DIVIDING FRACTIONS

The rules of arithmetic apply to the division of algebraic fractions; as in arithmetic, simply invert the divisor and multiply, as follows:

Arithmetic: $\dfrac{3}{8} \times \dfrac{9}{16} = \dfrac{3}{8} \times \dfrac{16}{9}$

$$= \dfrac{\cancel{3}}{\cancel{8}} \times \dfrac{\cancel{(8)}(2)}{\cancel{(3)}(3)} = \dfrac{2}{3}$$

Algebra: $\dfrac{x - 3y}{x + 3y} \div \dfrac{x^2 - 6xy + 9y^2}{x^2 + 7xy + 12y^2}$

$$= \dfrac{x - 3y}{x + 3y} \cdot \dfrac{x^2 + 7xy + 12y^2}{x^2 - 6xy + 9y^2}$$

$$= \dfrac{\cancel{x - 3y}}{\cancel{x + 3y}} \cdot \dfrac{\cancel{(x + 3y)}(x + 4y)}{\cancel{(x - 3y)}(x - 3y)}$$

$$= \dfrac{x + 4y}{x - 3y}$$

Practice problems. In the following problems, divide and reduce to lowest terms:

1. $\dfrac{x - 2}{x^2 + 4x + 4} \div \dfrac{1}{x^2 - 4}$

2. $\dfrac{2a - 1}{a^3 + 3a} \div \dfrac{a + 1}{a^2 + 3}$

3. $\dfrac{a^3 - 4a^2 + 3a}{a + 2} \div (a - 3)$

4. $\dfrac{6t + 12}{9t^2 + 6t - 24} \div \dfrac{8t - 12}{15t - 20}$

Answers:

1. $\dfrac{(x - 2)^2}{x + 2}$

3. $\dfrac{a(a - 1)}{a + 2}$

2. $\dfrac{2a - 1}{a^2 + a}$

4. $\dfrac{5}{4t - 6}$

ADDING AND SUBTRACTING FRACTIONS

The rules of arithmetic for adding and subtracting fractions are applicable to algebraic fractions. Fractions that are to be combined by addition or subtraction must have the same denominator. The numerators are then combined according to the operation indicated and the result is placed over the denominator. For example, in the expression

$$\frac{x - 4}{x - 2} + \frac{2 - 11x}{2 - x}$$

the second denominator will be the same as the first, if its sign is changed. The value of the fraction will remain the same if the sign of the numerator is also changed. Thus, we have the following simplification:

$$\frac{x - 4}{x - 2} + \frac{2 - 11x}{2 - x} = \frac{x - 4}{x - 2} + \frac{-(2 - 11x)}{-(2 - x)}$$

$$= \frac{x - 4}{x - 2} + \frac{11x - 2}{x - 2}$$

$$= \frac{x - 4 + 11x - 2}{x - 2}$$

$$= \frac{12x - 6}{x - 2}$$

$$= \frac{6(2x - 1)}{x - 2}$$

When the denominators are not the same, we must reduce all fractions to be added or subtracted to a common denominator and then proceed.

Consider, for example,

$$\frac{4}{x^2 - 4} + \frac{3}{x^2 - 4x - 12}$$

We first must find the least common denominator (LCD). Remember this is the least number that is exactly divisible by each of the denominators. To find such a number, as in arithmetic, we first separate each of the denominators into prime factors. The LCD will contain all of the various prime factors, each one as many times as it occurs in any of the denominators.

Factoring, we have

$$\frac{4}{(x + 2)(x - 2)} + \frac{3}{(x - 6)(x + 2)}$$

and the LCD is $(x + 2)(x - 2)(x - 6)$. Rewriting the fractions with this denominator and adding numerators, we have the following expression:

$$\frac{4(x - 6)}{(x + 2)(x - 2)(x - 6)} + \frac{3(x - 2)}{(x + 2)(x - 2)(x - 6)}$$

$$= \frac{4(x - 6) + 3(x - 2)}{\text{LCD}}$$

$$= \frac{4x - 24 + 3x - 6}{\text{LCD}}$$

$$= \frac{7x - 30}{(x + 2)(x - 2)(x - 6)}$$

As another example, consider

$$\frac{4}{x + 3} - \frac{x + 2}{x^2 + 4x + 3}$$

Factoring the denominator of the second fraction, we find that the LCD is $(x + 3)(x + 1)$. Rewriting the original fractions with the LCD as denominator, we may now combine the fractions as follows:

$$\frac{4(x + 1)}{(x + 3)(x + 1)} - \frac{(x + 2)}{(x + 3)(x + 1)}$$

$$= \frac{4x + 4 - x - 2}{(x + 3)(x + 1)}$$

$$= \frac{3x + 2}{(x + 3)(x + 1)}$$

Practice problems. Perform the indicated operations in each of the following problems:

1. $\dfrac{3x - 4}{x^2 + x - 2} - \dfrac{x - 2}{x - 1}$

2. $\dfrac{3a}{a^2 - 9} - \dfrac{3}{3 - a}$

3. $\dfrac{x - 3}{3x} + \dfrac{x + 2}{2x}$

4. $\dfrac{1}{a^4 - 1} - \dfrac{1}{a + 1}$

5. $\dfrac{3}{(a + 4)^2} - \dfrac{2}{a(a + 4)} + \dfrac{1}{6(a + 4)}$

Answers:

1. $\dfrac{3x - x^2}{(x + 2)(x - 1)}$

2. $\dfrac{6a + 9}{(a + 3)(a - 3)}$

3. $\dfrac{5}{6}$

4. $\dfrac{2 - a^3 + a^2 - a}{(a^2 + 1)(a + 1)(a - 1)}$

5. $\dfrac{a^2 + 10a - 48}{6a(a + 4)^2}$

CHAPTER 11

LINEAR EQUATIONS IN ONE VARIABLE

One of the principal reasons for an intensive study of polynomials, grouping symbols, factoring, and fractions is to prepare for solving equations. The equation is perhaps the most important tool in algebra, and the more skillful the student becomes in working with equations, the greater will be his ease in solving problems.

Before learning to solve equations, it is necessary to become familiar with the words used in the discussion of them. An EQUATION is a statement that two expressions are equal in value. Thus,

$$4 + 5 = 9$$

and

$$A = lw$$

(Area of a rectangle = length x width)

are equations. The part to the left of the equality sign is called the LEFT MEMBER, or first member, of the equation. The part to the right is the RIGHT MEMBER, or second member, of the equation.

The members of an equation are sometimes thought of as corresponding to two weights that balance a scale. (See fig. 11-1.) This comparison is often helpful to students who are learning to solve equations. It is obvious, in

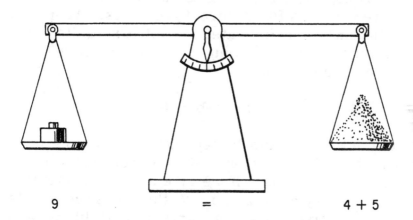

9   =   4 + 5

Figure 11-1. Equation compared to a
balance scale.

the case of the scale, that any change made in one pan must be accompanied by an equal change in the other pan. Otherwise the scale will not balance. Operations on equations are based on the same principle. The members must be kept balanced or the equality is lost.

CONSTANTS AND VARIABLES

Expressions in algebra consist of constants and variables. A CONSTANT is a quantity whose value remains the same throughout a particular problem. A VARIABLE is a quantity whose value is free to vary.

There are two kinds of constants--fixed and arbitrary. Numbers such as 7, -3, 1/2, and π are examples of FIXED constants. Their values never change. In $5x + 7 = 0$, the numbers 5 and 7 are fixed constants.

ARBITRARY constants can be assigned different values for different problems. Arbitrary constants are indicated by letters—quite often letters at the beginning of the alphabet such as a, b, c, and d. In

$$ax + b = 0,$$

the letters a and b represent arbitrary constants. The form $ax + b = 0$ represent many linear equations. If we give a and b particular values, say $a = 5$ and $b = 7$, then these constants become fixed, for this particular problem, and the equation becomes

$$5x + 7 = 0$$

A variable may have one value or it may have many values in a discussion. The letters at the end of the alphabet, such as x, y, z, and w, usually are used to represent variables. In $5x + 7$, the letter x is the variable. If $x = 1$, then

$$5x + 7 = 5 + 7 = 12$$

If $x = 2$, then

$$5x + 7 = 5(2) + 7 = 10 + 7 = 17$$

and so on for as many values of x as we desire to select.

If the expression 5x + 7 is set equal to some particular number, say -23, then the resulting equality

$$5x + 7 = -23$$

holds true for just one value of x. The value is -6, since

$$5(-6) + 7 = -23$$

In an algebraic expression, terms that contain a variable are called VARIABLE TERMS. Terms that do not contain a variable are CONSTANT TERMS. The expression 5x + 7 contains one variable term and one constant term. The variable term is 5x, while 7 is the constant term. In ax + b, ax is the variable term and b is the constant term.

A variable term often is designated by naming the variable it contains. In 5x + 7, 5x is the x-term. In ax + by, ax is the x-term, while by is the y-term.

DEGREE OF AN EQUATION

The degree of an equation that has not more than one variable in each term is the exponent of the highest power to which that variable is raised in the equation. The equation

$$3x - 17 = 0$$

is a FIRST-DEGREE equation, since x is raised

only to the first power.

An example of a SECOND-DEGREE equation is

$$5x^2 - 2x + 1 = 0.$$

The equation,

$$4x^3 - 7x^2 = 0,$$

is of the THIRD DEGREE.

The equation,

$$3x - 2y = 5$$

is of the first degree in two variables, x and y. When more than one variable appears in a term, as in $xy = 5$, it is necessary to add the exponents of the variables within a term to get the degree of the equation. Since $1 + 1 = 2$, the equation $xy = 5$ is of the second degree.

LINEAR EQUATIONS

Graphs are used in many different forms to give visual pictures of certain related facts. For example, they are used to show business trends, production output, continued individual attainment, and so forth. We find bar graphs, line graphs, circle graphs, and many other types, each of which is used for a particular need. In algebra, graphs are also used to give a visual picture containing a great deal of information about equations.

Sometimes many numerical values, when

substituted for the variables of an equation, will satisfy the conditions of the equation. On a particular type of graph (which will be explained fully in chapter 12) several of these values are plotted (located), and when enough are plotted, a line is drawn through these points. For each particular equation a certain type of curve results. For equations in the first degree in one or two variables, the resulting shape of the "curve" is a straight line. Thus, the name LINEAR EQUATION is derived. Equations of a higher degree form various other shapes. The name "linear equation" now applies to equations of the first degree, regardless of the number of variables they contain. Chapter 12 shows how an equation may be pictured on a graph. The purpose and value of graphing an equation will also be developed.

IDENTITIES

If a statement of equality involves one or more variables, it may be either an IDENTITY (identical equation) or a CONDITIONAL EQUATION. An identity is an equality that states a fact, such as the following examples:

1. $9 + 5 = 14$

2. $2n + 5n = 7n$

3. $6(x - 3) = 6x - 18$

Notice that equation 3 merely shows the factored form of $6x - 18$ and holds true when any value of x is substituted. For example, if $x = 5$, it becomes

$$6(5-3) = 6(5) - 18$$

$$6(2) = 30 - 18$$

$$12 = 12$$

If x assumes the negative value -10, this identity becomes

$$6(-10-3) = 6(-10)-18$$

$$6(-13) = -60-18$$

$$-78 = -78$$

An identity is established when both sides of
the equality have been reduced to the same
number or the same expression. When 5 is
substituted for x, the value of either side of
$6(x-3) = 6x - 18$ is 12. When -10 is substituted
for x, the value on either side is -78. The fact
that this equality is an identity can be shown
also by factoring the right side so that the
equality becomes

$$6(x-3) = 6(x-3)$$

The expressions on the two sides of the equality
are identical.

CONDITIONAL EQUATIONS

A statement such as $2x-1 = 0$ is an equality
only when x has one particular value. Such a
statement is called a CONDITIONAL EQUA-
TION, since it is true only under the condition
that $x = 1/2$. Likewise, the equation $y - 7 = 8$

320

holds true only if y = 15.

The value of the variable for which an equation in one variable holds true is a ROOT, or SOLUTION, of the equation. When we speak of solving equations in algebra, we refer to conditional equations. The solution of a conditional equation can be verified by substituting for the variable its value, as determined by the solution.

The solution is correct if the equality reduces to an identity. For example, if 1/2 is substituted for x in 2x - 1 = 0, the result is

$$2\left(\frac{1}{2}\right) - 1 = 0$$

$$1 - 1 = 0$$

$$0 = 0 \quad \text{(an identity)}$$

The identity is established for $x = \frac{1}{2}$, since the value of each side of the equality reduces to zero.

SOLVING LINEAR EQUATIONS

Solving a linear equation in one variable means finding the value of the variable that makes the equation true. For example, 11 is the SOLUTION of x - 7 = 4, since 11 - 7 = 4. The number 11 is said to SATISFY the equation. Basically, the operation used in solving equations is to manipulate both members, by addition, subtraction, multiplication, or division until the value of the variable becomes apparent. This manipulation may be accomplished in

321

a straightforward manner by use of the axioms outlined in chapter 3 of this course. These axioms may be summed up in the following rule: If both members of an equation are increased, decreased, multiplied, or divided by the same number, or by equal numbers, the results will be equal. (Division by zero is excluded.)

As mentioned earlier, an equation may be compared to a balance. What is done to one member must also be done to the other to maintain a balance. An equation must always be kept in balance or the equality is lost. We use the above rule to remove or adjust terms and coefficients until the value of the variable is discovered. Some examples of equations solved by means of the four operations mentioned in the rule are given in the following paragraphs.

ADDITION

Find the value of x in the equation

$$x - 3 = 12$$

As in any equation, we must isolate the variable on either the right or left side. In this problem, we leave the variable on the left and perform the following steps:

1. Add 3 to both members of the equation, as follows:

$$x - 3 + 3 = 12 + 3$$

In effect, we are "undoing" the subtraction indicated by the expression x - 3, for the purpose of isolating x in the left member.

2. Combining terms, we have

$$x = 15$$

SUBTRACTION

Find the value of x in the equation

$$x + 14 = 24$$

1. Subtract 14 from each member. In effect, this undoes the addition indicated in the expression x + 14.

$$x + 14 - 14 = 24 - 14$$

2. Combining terms, we have

$$x = 10$$

MULTIPLICATION

Find the value of y in the equation

$$\frac{y}{5} = 10$$

1. The only way to remove the 5 so that the y can be isolated is to undo the indicated division. Thus we use the inverse of division, which is multiplication. Multiplying both members by 5, we have the following: $5\left(\frac{y}{5}\right) = 5(10)$

2. Performing the indicated multiplications, we have

$$y = 50$$

DIVISION

Find the value of x in the equation

$$3x = 15$$

1. The multiplier 3 may be removed from the x by dividing the left member by 3. This must be balanced by dividing the right member by 3 also, as follows:

$$\frac{3x}{3} = \frac{15}{3}$$

2. Performing the indicated divisions, we have

$$x = 5$$

Practice problems. Solve the following equations:

1. m + 2 = 8   4. $\frac{x}{14} = 2$

2. x - 5 = 11   5. 2n = 5

3. 6x = -48   6. $\frac{1}{6}y = 6$

Answers:

1. m = 6   4. x = 28

2. x = 16   5. n = $2\frac{1}{2}$

3. x = -8   6. y = 36

SOLUTIONS REQUIRING MORE THAN ONE OPERATION

Most equations involve more steps in their solutions than the simple equations already described, but the basic operations remain unchanged. If the basic axioms are kept well in mind, these more complicated equations will not become too difficult. Equations may require one or all of the basic operations before a solution can be obtained.

Subtraction and Division

Find the value of x in the following equation:

$$2x + 4 = 16$$

1. The term containing x is isolated on the left by subtracting 4 from the left member. This operation must be balanced by also subtracting 4 from the right member, as follows:

$$2x + 4 - 4 = 16 - 4$$

2. Performing the indicated operations, we have

$$2x = 12$$

3. The multiplier 2 is removed from the x by dividing both sides of the equation by 2, as follows:

$$\frac{2x}{2} = \frac{12}{2}$$

$$x = 6$$

Addition, Multiplication, and Division

Find the value of y in the following equation:

$$\frac{3y}{2} - 4 = 11$$

1. Isolate the term containing y on the left by adding 4 to both sides, as follows:

$$\frac{3y}{2} - 4 + 4 = 11 + 4$$

$$\frac{3y}{2} = 15$$

2. Since the 2 will not divide the 3 exactly, multiply the left member by 2 in order to eliminate the fraction. This operation must be balanced by multiplying the right member by 2, as follows:

$$2\left(\frac{3y}{2}\right) = 2(15)$$

$$3y = 30$$

3. Divide both members by 3, in order to isolate the y in the left member, as follows:

$$\frac{3y}{3} = \frac{30}{3}$$

$$y = 10$$

Equations Having the Variable in More Than One Term

Find the value of x in the following equation:

$$\frac{3x}{4} + x = 12 - x$$

1. Rewrite the equation with no terms containing the variable in the right member. This requires adding x to the right member to eliminate the -x term, and balance requires that we also add x to the left member, as follows:

$$\frac{3x}{4} + x + x = 12 - x + x$$

$$\frac{3x}{4} + 2x = 12$$

2. Since the 4 will not divide the 3 exactly, it is necessary to multiply the first term by 4 to eliminate the fraction. However, notice that this multiplication cannot be performed on the first term only; any multiplier which is introduced for simplification purposes must be applied to the entire equation. Thus each term in the equation is multiplied by 4, as follows:

$$4\left(\frac{3x}{4}\right) + 4(2x) = 4(12)$$

$$3x + 8x = 48$$

3. Add the terms containing x and then divide both sides by 11 to isolate the x in the left member, as follows:

$$11x = 48$$

$$x = \frac{48}{11}$$

$$= 4\frac{4}{11}$$

Practice problems. Solve each of the following equations:

1. $x - 1 = \frac{1}{2}$ 4. $4 - 7x = 9 - 8x$

2. $\frac{y}{3} + y = 8$ 5. $\frac{y}{2} + 6y = 13$

3. $\frac{x}{4} + 3x = 7$ 6. $\frac{1}{2}x - 2x = 25 + x$

Answers:

1. $x = 3/2$ 4. $x = 5$

2. $y = 6$ 5. $y = 2$

3. $x = 28/13$ 6. $x = -10$

EQUATIONS WITH LITERAL COEFFICIENTS

As stated earlier, the first letters of the

alphabet usually represent known quantities (constants), and the last letters represent unknown quantities (variables). Thus, we usually solve for x, y, or z.

An equation such as

$$ax - 8 = bx - 5$$

has letters as coefficients. Equations with literal coefficients are solved in the same way as equations with numerical coefficients, except that when an operation cannot actually be performed, it merely is indicated.

In solving for x in the equation

$$ax - 8 = bx - 5$$

subtract bx from both members and add 8 to both members. The result is

$$ax - bx = 8 - 5$$

Since the subtraction on the left side cannot actually be performed, it is indicated. The quantity, a - b, is the coefficient of x when terms are collected. The equation takes the form

$$(a-b) \; x = 3$$

Now divide both sides of the equation by a-b. Again the result can be only indicated. The solution of the equation is

$$x = \frac{3}{a-b}$$

In solving for y in the equation

$$ay + b = 4$$

subtract b from both members as follows:

$$ay = 4 - b$$

Dividing both members by a, the solution is

$$y = \frac{4-b}{a}$$

Practice problems. Solve for x in each of the following:

1. $3 + x = b$

2. $4x = 8 + t$

3. $3x + 6m = 7m$

4. $ax - 2(x + b) = 3a$

Answers:

1. $x = b - 3$

2. $x = \frac{8 + t}{4}$

3. $x = \frac{m}{3}$

4. $x = \frac{3a + 2b}{a - 2}$

REMOVING SIGNS OF GROUPING

If signs of grouping appear in an equation they should be removed in the manner indicated in chapter 9 of this course. For example, solve the equation

$$5 = 24 - [x-12(x-2) - 6(x-2)]$$

Notice that the same expression, x-2, occurs in both parentheses. By combining the terms containing (x-2), the equation becomes

$$5 = 24 - [x-18(x-2)]$$

Next, remove the parentheses and then the bracket, obtaining

$$5 = 24 - [x-18x + 36]$$
$$= 24 - [36 - 17x]$$
$$= 24 - 36 + 17x$$
$$= -12 + 17x$$

Subtracting 17x from both members and then subtracting 5 from both members, we have

$$-17x = -12 - 5$$
$$-17x = -17$$

Divide both members by -17. The solution is

$$x = 1$$

EQUATIONS CONTAINING FRACTIONS

To solve for x in an equation such as

$$\frac{2x}{3} + \frac{x}{12} - 1 = \frac{1}{4} + \frac{x}{2}$$

first clear the equation of fractions. To do this, find the least common denominator of the fractions. Then multiply both sides of the equation by the LCD.

The least common denominator of 3, 12, 4, and 2 is 12. Multiply both sides of the equation by 12. The resulting equation is

$$8x + x - 12 = 3 + 6x$$

Subtract 6x from both members, add 12 to both members, and collect like terms as follows:

$$9x - 6x = 12 + 3$$

$$3x = 15$$

The solution is

$$x = 5$$

To prove that x = 5 is the correct solution, substitute 5 for x in the original equation and show that both sides of the equation reduce to the same value. The result of substitution is

$$\frac{2(5)}{3} + \frac{5}{12} - 1 = \frac{1}{4} + \frac{5}{2}$$

In establishing an identity, the two sides of the equality are treated separately, and the operations are performed as indicated. Sometimes, as here, fractions occur on both sides of the equality, and it is desirable to find the least common denominator for more than one set of fractions. The same denominator could be used on both sides of the equality, but this might make some of the terms of the fractions larger than necessary.

Proceeding in establishing the identity for

x = 5 in the foregoing equation we obtain

$$\frac{10}{3} + \frac{5}{12} - \frac{3}{3} = \frac{1}{4} + \frac{10}{4}$$

$$\frac{7}{3} + \frac{5}{12} = \frac{11}{4}$$

$$\frac{28}{12} + \frac{5}{12} = \frac{11}{4}$$

$$\frac{33}{12} = \frac{11}{4}$$

$$\frac{11}{4} = \frac{11}{4}$$

Each member of the equality has the value 11/4 when x = 5. The fact that the equation becomes an identity when x is replaced by 5 proves that x = 5 is the solution.

Practice problems. Solve each of the following equations:

1. $\frac{x}{4} - 2 = \frac{x}{6}$

3. $\frac{y}{2} - \frac{y}{3} = 5$

2. $\frac{1}{2} - \frac{1}{v} = \frac{1}{3}$

4. $\frac{3}{4x} = 6$

Answers:

1. x = 24

3. y = 30

2. v = 6

4. x = 1/8

GENERAL FORM OF A LINEAR EQUATION

The expression GENERAL FORM, in mathematics, implies a form to which all expressions or equations of a certain type can be reduced. The only possible terms in a linear equation in one variable are the first-degree term and the constant term. Therefore, the general form of a linear equation in one variable is

$$ax + b = 0$$

By selecting various values for a and b, this form can represent any linear equation in one variable after such an equation has been simplified. For example, if $a = 7$ and $b = 5$, $ax + b = 0$ represents the numerical equation

$$7x + 5 = 0$$

If $a = 2m - n$ and $b = p - q$, then $ax + b = 0$ represents the literal equation

$$(2m-n)x + p - q = 0$$

This equation is solved as follows:

$$(2m-n)x + (p-q) - (p-q) = 0 - (p-q)$$
$$(2m-n)x = 0 - (p-q)$$

$$x = \frac{-(p-q)}{2m-n} = \frac{q-p}{2m-n}$$

USING EQUATIONS TO SOLVE PROBLEMS

To solve a problem, we first translate the

numerical sense of the problem into an equation. To see how this is accomplished, consider the following examples and their solutions.

EXAMPLE 1: Together Smith and Jones have $120. Jones has 5 times as much as Smith. How much has Smith?

SOLUTION:

Step 1. Get the problem clearly in mind. There are two parts to each problem—what is given (the facts) and what we want to know (the question). In this problem we know that Jones has 5 times as much as Smith and together they have $120. We want to know how much Smith has.

Step 2. Express the unknown as a letter. Usually we express the unknown or number we know the least about as a letter (conventionally we use x). Here we know the least about Smith's money. Let x represent the number of dollars Smith has.

Step 3. Express the other facts in terms of the unknown. If x is the number of dollars Smith has and Jones has 5 times as much, then 5x is the number of dollars Jones has.

Step 4. Express the facts as an equation. The problem will express or imply a relation between the expressions in steps 2 and 3. Smith's dollars plus Jones' dollars equal $120. Translating this statement into algebraic symbols, we have

$$x + 5x = 120$$

Solving the equation for x,

$$6x = 120$$

$$x = 20$$

Thus Smith has $20.

Step 5. Check: See if the solution satisfies the original statement of the problem. Smith and Jones have $120.

$20 + $100 = $120
(Smith's money) (Jones' money)

EXAMPLE 2: Brown can do a piece of work in 5 hr. If Olsen can do it in 4 hr how long will it take them to do the work together?

SOLUTION:

Step 1. Given: Brown could do the work in 5 hr. Olsen could do it in 4 hours.

Unknown: How long it takes them to do the work together.

Step 2. Let x represent the time it takes them to do the work together.

Step 3. Then $\frac{1}{x}$ is the amount they do together in 1 hr. Also, in 1 hour Brown does $\frac{1}{5}$ of the work and Olsen does $\frac{1}{4}$ of the work.

Step 4. The amount done in 1 hr is equal to the part of the work done by Brown in 1 hr plus that done by Olsen in 1 hr.

$$\frac{1}{x} = \frac{1}{5} + \frac{1}{4}$$

Solving the equation,

$$20x\left(\frac{1}{x}\right) = 20x\left(\frac{1}{5}\right) + 20x\left(\frac{1}{4}\right)$$

$$20 = 4x + 5x$$

$$20 = 9x$$

$$\frac{20}{9} = x, \text{ or } x = 2\frac{2}{9} \text{ hours}$$

They complete the work together in $2\frac{2}{9}$ hours.

Step 5. Check: $2\frac{2}{9} \times \frac{1}{5}$ = amount Brown does

$$2\frac{2}{9} \times \frac{1}{4} = \text{amount Olsen does}$$

$$\left(\frac{20}{9} \times \frac{1}{5}\right) + \left(\frac{20}{9} \times \frac{1}{4}\right) = \frac{4}{9} + \frac{5}{9} = \frac{9}{9}$$

Practice problems. Use a linear equation in one variable to solve each of the following problems:

1. Find three numbers such that the second is twice the first and the third is three times as large as the first. Their sum is 180.

2. A seaman drew $75.00 pay in dollar bills and five-dollar bills. The number of dollar bills was three more than the number of five-dollar bills. How many of each kind did he draw? (Hint: If x is the number of five-dollar

bills, then 5x is the number of dollars they represent.)

3. Airman A can complete a maintenance task in 4 hr. Airman B requires only 3 hr to do the same work. If they work together, how long should it take them to complete the job?

Answers:

1. First number is 30.
    Second number is 60.
    Third number is 90.

2. Number of five-dollar bills is 12.
    Number of one-dollar bills is 15.

3. $1\frac{5}{7}$ hr.

INEQUALITIES

Modern mathematical thought gives considerable emphasis to the concept of inequality. A meaningful comparison between two quantities can be set up if they are related in some way, even though the relationship may not be one of equality.

The expression "number sentence" is often used to describe a general relationship which may be either an equality or an inequality. If the number sentence states an equality, it is an EQUATION; if it states an inequality, it is an INEQUATION.

ORDER PROPERTIES OF REAL NUMBERS

The idea of order, or relative rank according to size, is based upon two intuitive concepts: "greater than" and "less than." Mathematicians use the symbol $>$ to represent "greater than" and the symbol $<$ to represent "less than." For example, the inequation stating that 7 is greater than 5 is written in symbols as follows:

$$7 > 5$$

The inequation stating that x is less than 10 is written as follows:

$$x < 10$$

A "solution" of an inequation involving a variable is any number which may be substituted for the variable without changing the relationship between the left member and the right member. For example, the inequation $x < 10$ has many solutions. All negative numbers, and all positive numbers between 0 and 9, may be substituted for x successfully. These solutions comprise a set of numbers, called the SOLUTION SET.

The SENSE of an inequality refers to the direction in which the inequality symbol points. For example, the following two inequalities have opposite sense:

$$7 > 5$$

$$10 < 12$$

PROPERTIES OF INEQUALITIES

Inequations may be manipulated in accordance with specific operational rules, in a manner similar to that used with equations.

Addition

The rule for addition is as follows: If the same quantity is added to both members of an inequation, the result is an inequation having the same sense as the original inequation. The following examples illustrate this:

1. $5 < 8$

$5 + 2 < 8 + 2$

$7 < 10$

The addition of 2 to both members does not change the sense of the inequation.

2. $5 < 8$

$5 + (-3) < 8 + (-3)$

$2 < 5$

The addition of -3 to both members does not change the sense of the inequation.

Addition of the same quantity to both members is a useful method for solving inequations. In the following example, 2 is added to both members in order to isolate the x term on the left: $x - 2 > 6$

$x - 2 + 2 > 6 + 2$

$x > 8$

Multiplication

The rule for multiplication is as follows: If both members of an inequation are multiplied by the same positive quantity, the sense of the resulting inequation is the same as that of the original inequation. This is illustrated as follows:

$$1. \quad -3 < -2$$

$$2(-3) < 2(-2)$$

$$-6 < -4$$

Multiplication of both members by 2 does not change the sense of th inequation.

$$2. \quad 10 < 12$$

$$\frac{1}{2}(10) < \frac{1}{2}(12)$$

$$5 < 6$$

Multiplication of both members by 1/2 does not change the sense of the inequation.

Notice that example 2 illustrates division of both members by 2. Since any division can be rewritten as multiplication by a fraction, the multiplication rule is applicable to both multiplication and division.

Multiplication is used to simplify the solution of inequations such as the following:

$$\frac{x}{3} > 2$$

Multiply both members by 3:

$$3\left(\frac{x}{3}\right) > 3(2)$$

$$x > 6$$

Sense Reversal

If both sides of an inequation are multiplied or divided by the same negative number, the sense of the resulting inequation is reversed. This is illustrated as follows:

1. $-4 < -2$

$$(-3)(-4) > (-3)(-2)$$

$$12 > 6$$

2. $7 > 5$

$$(-2)(7) < (-2)(5)$$

$$-14 < -10$$

Sense reversal is useful in the solution of an inequation in which the variable is preceded by a negative sign, as follows:

$$2 - x < 4$$

Add -2 to both members to isolate the x term:

$$2 - x - 2 < 4 - 2$$

$$- x < 2$$

Multiply both members by -1:

$$x > -2$$

Practice problems. Solve each of the following inequations:

1. $x + 2 > 3$ 3. $3 - x < 6$

2. $\dfrac{y}{3} - 1 < 2$ 4. $4y > 8$

Answers:

1. $x > 1$ 3. $x > -3$

2. $y < 9$ 4. $y > 2$

GRAPHING INEQUALITIES

An inequation such as $x > 2$ can be graphed on a number line, as shown in figure 11-2.

The heavy line in figure 11-2 contains all values of x which comprise the solution set. Notice that this line continues indefinitely in the positive direction, as indicated by the arrow head. Notice also that the point representing $x = 2$ is designated by a circle. This signifies that the solution set does not contain the number 2.

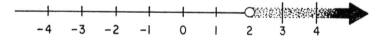

Figure 11-2.—Graph of the inequation $x > 2$.

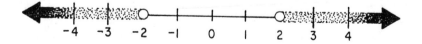

Figure 11-3.—Graph of $x^2 > 4$.

343

Figure 11-3 is a graph of the inequation $x^2 > 4$. Since the square of any number greater than 2 is greater than 4, the solution set contains all values of x greater than 2. Furthermore, the solution set contains all values of x less than -2. This is because the square of any negative number smaller than -2 is a positive number greater than 4.

CHAPTER 12

LINEAR EQUATIONS IN TWO VARIABLES

Thus far in this course, discussions of equations have been limited to linear equations in one variable. Linear equations which have two variables are common, and their solution involves extending some of the procedures which have already been introduced.

RECTANGULAR COORDINATES

An outstanding characteristic of equations in two variables is their adaptability to graphical analysis. The rectangular coordinate system, which was introduced in chapter 3 of this course, is used in analyzing equations graphically. This system of vertical and horizontal lines, meeting each other at right angles and thus forming a rectangular grid, is often called the Cartesian coordinate system. It is named after the French philosopher and mathematician, Rene Descartes, who invented it.

COORDINATE AXES

The rectangular coordinate system is developed on a framework of reference similar to figure 3-2 in chapter 3 of this course. On a piece of graph paper, two lines are drawn intersecting each other at right angles, as in figure 12-1. The vertical line is usually labeled with the capital letter Y and called the Y axis. The horizontal line is usually labeled with the capital letter X and called the X axis. The point where the X and Y axes intersect is called the ORIGIN and is labeled with the letter o.

Above the origin, numbers measured along or parallel to the Y axis are positive; below the origin they are negative. To the right of the origin, numbers measured along or parallel to the X axis are positive; to the left they are negative.

COORDINATES

A point anywhere on the graph may be located by two numbers, one showing the distance of the point from the Y axis, and the other showing the distance of the point from the X axis. Point P (fig. 12-1) is 6 units to the right of the Y axis and 3 units above the X axis. We call the numbers that indicate the position of a point COORDINATES. The number indicating the distance of the point measured horizontally from the origin is the X coordinate (6 in this example), and the number indicating the distance of the point measured vertically from the origin (3 in this example) is the Y coordinate.

In describing the location of a point by means of rectangular coordinates, it is customary to place the coordinates within parentheses and separate them with a comma. The X coordinate is always written first. The coordinates of point P (fig. 12-1) are written (6, 3). The coordinates for point Q are (4, -5); for point R, they are (-5, -2); and for point S, they are (-8, 5).

Usually when we indicate a point on a graph, we write a letter and the coordinates of the point. Thus, in figure 12-1, for point S, we write S(-8, 5). The other points would ordinarily be written, P(6, 3), Q(4, -5), and R(-5, -2). The Y coordinate of a point is often called its ORDINATE and the X coordinate is often called its ABSCISSA.

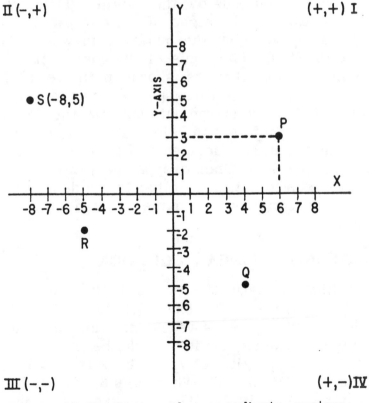

Figure 12-1.—Rectangular coordinate system.

QUADRANTS

The X and Y axes divide the graph into four parts called QUADRANTS. In figure 12-1, point P is in quadrant I, point S is in quadrant II, R is in quadrant III, and Q is in quadrant IV. In the first and fourth quadrants, the X coordinate is positive, because it is to the right of the origin. In the second and third quadrant it is negative, because it is to the left of the origin. Likewise, the Y coordinate is positive in the first and second quadrants, being above the origin; it is negative in the third and fourth quadrants, being below the origin. Thus, we know in advance the signs of the coordinates of a point by knowing the quadrant in which the point appears. The signs of the coordinates in the four quadrants are shown in figure 12-1.

Locating points with respect to axes is called PLOTTING. As shown with point P (fig. 12-1), plotting a point is equivalent to completing a rectangle that has segments of the axes as two of its sides with lines dropped perpendicularly to the axes forming the other two sides. This is the reason for the name "rectangular co-ordinates."

PLOTTING A LINEAR EQUATION

A linear equation in two variables may have many solutions. For example, in solving the equation $2x - y = 5$, we can find an unlimited number of values of x for which there will be a corresponding value of y. When x is 4, y is 3, since $(2 \times 4) - 3 = 5$. When x is 3, y is 1, and when x is 6, y is 7. When we graph an equation, these pairs of values are considered co-ordinates of points on the graph. The graph of an equation is nothing more than a line joining the points located by the various pairs of numbers that satisfy the equation.

To picture an equation, we first find several pairs of values that satisfy the equation. For example, for the equation 2x - y = 5, we assign several values to x and solve for y. A convenient way to find values is to first solve the equation for either variable, as follows:

$$2x - y = 5$$

$$-y = -2x + 5$$

$$y = 2x - 5$$

Once this is accomplished, the value of y is readily apparent when values are substituted for x. The information derived may be recorded in a table such as table 12-1. We then lay off X and Y axes on graph paper, select some convenient unit distance for measurement along the axes, and then plot the pairs of values found for x and y as coordinates of points on the graph. Thus, we locate the pairs of values shown in table 12-1 on a graph, as shown in figure 12-2 (A).

Table 12-1.—Values of x and y in the equation 2x - y = 5.

If x = ------	-2	1	3	5	6	7	8
Then y = ---	-9	-3	1	5	7	9	11

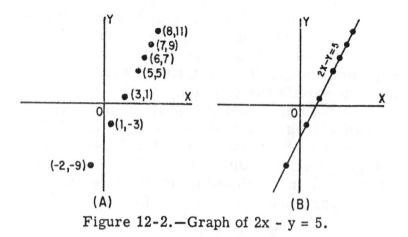

Figure 12-2.—Graph of 2x - y = 5.

Finally, we draw a line joining these points, as in figure 12-2 (B). It is seen that this is a straight line; hence the name "linear equation." Once the graph is drawn, it is customary to write the equation it represents along the line, as shown in figure 12-2 (B).

It can be shown that the graph of an equation is the geometric representation of all the points whose coordinates satisfy the conditions of the equation. The line represents an infinite number of pairs of coordinates for this equation. For example, selecting at random the point on the line where x is $2\frac{1}{2}$ and y is 0 and substituting these values in the equation, we find that they satisfy it. Thus,

$$2\left(2\frac{1}{2}\right) - 0 = 5$$

If two points that lie on a straight line can be located, the position of the line is known. The mathematical language for this is "Two points DETERMINE a straight line." We now

350

know that the graph of a linear equation in two variables is a straight line. Since two points are sufficient to determine a straight line, a linear equation can be graphed by plotting two points and drawing a straight line through these points. Very often pairs of whole numbers which satisfy the equation can be found by inspection. Such points are easily plotted.

After the line is drawn through two points, it is well to plot a third point as a check. If this third point whose coordinates satisfy the equation lies on the line the graph is accurately drawn.

X AND Y INTERCEPTS

Any straight line which is not parallel to one of the axes has an X intercept and a Y intercept. These are the points at which the line crosses the X and Y axes. At the X intercept, the graph line is touching the X axis, and thus the Y value at that point is 0. At the Y intercept, the graph line is touching the Y axis; the X value at that point is 0.

In order to find the X intercept, we simply let y = 0 and find the corresponding value of x. The Y intercept is found by letting x = 0 and finding the corresponding value of y. For example, the line

$$5x + 3y = 15$$

crosses the Y axis at (0,5). This may be verified by letting x = 0 in the equation. The X intercept is (3,0), since x is 3 when y is 0. Figure 12-3 shows the line

$$5x + 3y = 15$$

graphed by means of the X and Y intercepts.

EQUATIONS IN ONE VARIABLE

An equation containing only one variable is easily graphed, since the line it represents lies parallel to an axis. For example, in

$$2y = 9$$

the value of y is

$$\frac{9}{2}, \text{ or } 4\frac{1}{2}$$

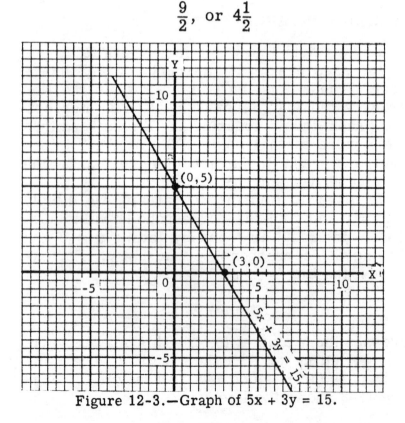

Figure 12-3.—Graph of 5x + 3y = 15.

The line 2y = 9 lies parallel to the X axis at a distance of $4\frac{1}{2}$ units above it. (See fig. 12-4.) Notice that each small division on the graph paper in figure 12-4 represents one-half unit.

The line 4x + 15 = 0 lies parallel to the Y axis. The value of x is $-\frac{15}{4}$. Since this value is negative, the line lies to the left of the Y axis at a distance of $3\frac{3}{4}$ units. (See fig. 12-4.)

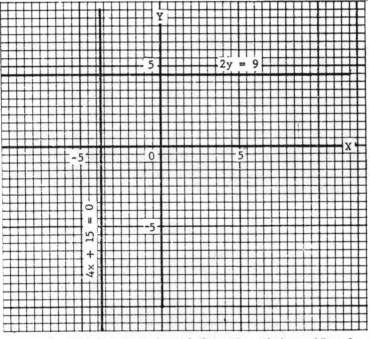

Figure 12-4.—Graphs of 2y = 9 and 4x + 15 = 0.

From the foregoing discussion, we arrive at two important conclusions:

1. A pair of numbers that satisfy an equation are the coordinates of a point on the graph

of the equation.

2. The coordinates of any point on the graph of an equation will satisfy that equation.

SOLVING EQUATIONS IN TWO VARIABLES

A solution of a linear equation in two variables consists of a pair of numbers that satisfy the equation. For example, x = 2 and y = 1 constitute a solution of

$$3x - 5y = 1$$

When 2 is substituted for x and 1 is substituted for y, we have

$$3(2) - 5(1) = 1$$

The numbers x = -3 and y = -2 also form a solution. This is true because substituting -3 for x and -2 for y reduces the equation to an identity:

$$3(-3) -5(-2) = 1$$
$$-9 + 10 = 1$$
$$1 = 1$$

Each pair of numbers (x, y) such as (2, 1) or (-3, -2) locates a point on the line 3x - 5y = 1. Many more solutions could be found. Any two numbers that constitute a solution of the equation are the coordinates of a point on the line represented by the equation.

Suppose we were asked to solve a problem

such as: Find two numbers such that their sum is 33 and their difference is 5. We could indicate the problem algebraically by letting x represent one number and y the other. Thus, the problem may be indicated by the two equations

$$x + y = 33$$
$$x - y = 5$$

Considered separately, each of these equations represents a straight line on a graph. There are many pairs of values for x and y which satisfy the first equation, and many other pairs which satisfy the second equation. Our problem is to find ONE pair of values that will satisfy BOTH equations. Such a pair of values is said to satisfy both equations at the same time, or simultaneously. Hence, two equations for which we seek a common solution are called SIMULTANEOUS EQUATIONS. The two equations, taken together, comprise a SYSTEM of equations.

Graphical Solution

If there is a pair of numbers which may be substituted for x and y in two different equations, these numbers form the coordinates of a point which lies on the graph of each equation. The only way in which a point can lie on two lines simultaneously is for the point to be at the intersection of the lines. Therefore, the graphical solution of two simultaneous equations involves drawing their graphs and locating the point at which the graph lines intersect.

For example, when we graph the equations $x + y = 33$ and $x - y = 5$, as in figure 12-5, we see that they intersect in a single point. There is one pair of values comprising coordinates of that point (19, 14), and that pair of values satisfies both equations, as follows:

$$x + y = 33 \qquad x - y = 5$$
$$19 + 14 = 33 \qquad 19 - 14 = 5$$

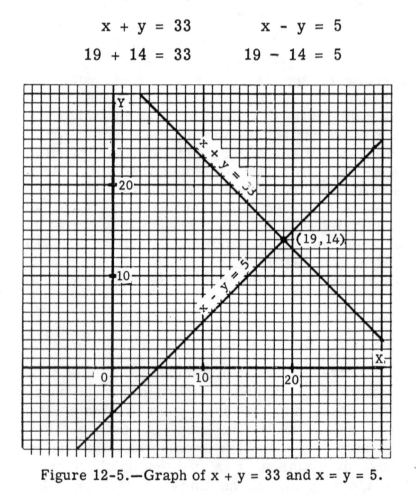

Figure 12-5.—Graph of $x + y = 33$ and $x = y = 5$.

This pair of numbers satisfies each equation. It is the only pair of numbers that satisfies the two equations simultaneously.

The graphical method is a quick and simple means of finding an approximate solution of two simultaneous equations. Each equation is graphed, and the point of intersection of the two lines is read as accurately as possible. A high degree of accuracy can be obtained but this, of course, is dependent on the precision with which the lines are graphed and the amount of accuracy possible in reading the graph. Sometimes the graphical method is quite adequate for the purpose of the problem.

Figure 12-6 shows the graphs of $x + y = 11$ and $x - y = -3$. The intersection appears to be the point $(4, 7)$. Substituting $x = 4$ and $y = 7$ into the equations shows that this is the actual point of intersection, since this pair of numbers satisfies both equations.

The equations $7x - 8y = 2$ and $4x + 3y = 5$ are graphed in figure 12-7. The lines intersect where y is approximately $1/2$ and x is approximately $5/6$.

Practice problems. Solve the following simultaneous systems graphically:

1. $x + y = 8$
   $x - y = 2$

2. $3x + 2y = 12$
   $4x + 5y = 2$

Answers:

1. $x = 5$
   $y = 3$

2. $x = 8$
   $y = -6$

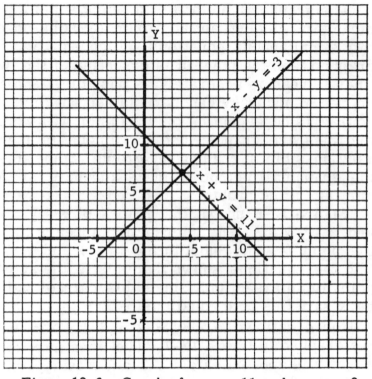

Figure 12-6.—Graph of x + y = 11 and x - y = -3.

Addition Method

The addition method of solving systems of equations is illustrated in the following example:

$$x - y = 2$$
$$x + y = 8$$
$$\overline{2x + 0 = 10}$$
$$x = 5$$

The result in the foregoing example is obtained

by adding the left member of the first equation to the left member of the second, and adding the right member of the first equation to the right member of the second.

Having found the value of x, we substitute this value in either of the original equations to find the value of y, as follows:

$$x - y = 2$$
$$(5) - y = 2$$
$$-y = 2 - 5$$
$$-y = -3$$
$$y = 3$$

Notice that the primary goal in the addition method is the elimination (temporarily) of one of the variables. If the coefficient of y is the same in both equations, except for its sign, adding the equations eliminates y as in the foregoing example. On the other hand, suppose that the coefficient of the variable which we desire to eliminate is exactly the same in both equations.

In the following example, the coefficient of x is the same in both equations, including its sign:

$$x + 2y = 4$$
$$x - 3y = -1$$

Adding the equations would not eliminate either x or y. However, if we multiply both members of the second equation by -1, then addition will eliminate x, as follows:

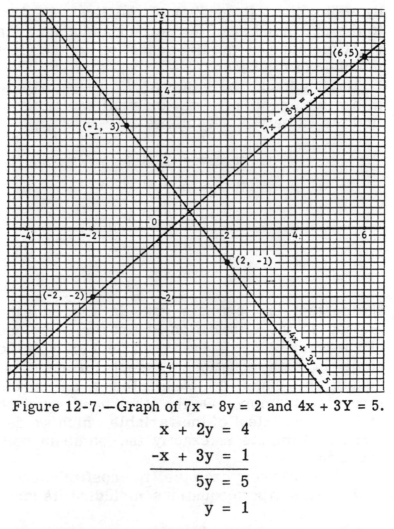

Figure 12-7.—Graph of 7x - 8y = 2 and 4x + 3Y = 5.

$$x + 2y = 4$$

$$\underline{-x + 3y = 1}$$

$$5y = 5$$

$$y = 1$$

The value of x is found by substituting 1 for y in either of the original equations, as follows:

$$x + 2(1) = 4$$

$$x = 2$$

As a second example of the addition method, find the solution of the simultaneous equations

$$3x + 2y = 12$$
$$4x + 5y = 2$$

Here both x and y have unlike coefficients. The coefficients of one of the variables must be made the same, except for their signs.

The coefficients of x will be the same except for signs, if both members of the first equation are multiplied by 4 and both members of the second equation by -3. Then addition will eliminate x.

Following this procedure to get the value of y, we multiply the first equation by 4 and the second equation by -3, as follows:

$$12x + 8y = 48$$
$$-12x - 15y = -6$$
$$\overline{-7y = 42}$$
$$y = -6$$

Substituting for y in the first equation to get the value of x, we have

$$3x + 2(-6) = 12$$
$$x + 2(-2) = 4$$
$$x - 4 = 4$$
$$x = 8$$

This solution is checked algebraically by substituting 8 for x and -6 for y in each of the original equations, as follows:

1. $3x + 2y = 12$

$3(8) + 2(-6) = 12$

$24 - 12 = 12$

2. $4x + 5y = 2$

$4(8) + 5(-6) = 2$

$32 - 30 = 2$

Practice problems. Use the addition method to solve the following problems:

1. $x + y = 24$

   $x - y = 12$

2. $5t + 2v = 9$

   $3t - 2v = -5$

3. $x - 2y = -1$

   $2x + 3y = 12$

4. $2x + 7y = 3$

   $3x - 5y = 51$

Answers:

1. x = 18

   y = 6

2. t = 1/2

   $v = \dfrac{13}{4}$

3. x = 3

   y = 2

4. x = 12

   y = -3

Substitution Method

In some cases it is more convenient to use the substitution method of solving problems. In this method we solve one equation for one of the variables and substitute the value obtained into the other equation. This eliminates one of the variables, leaving an equation in one unknown. For example, find the solution of the following system:

$$4x + y = 11$$
$$x + 2y = 8$$

It is easy to solve for either y in the first equation or x in the second equation. Let us solve for v in the first equation. The result is

$$y = 11 - 4x$$

Since equals may be substituted for equals, we may substitute this value of y wherever **y** appears in the second equation. Thus,

$$x + 2(11 - 4x) = 8$$

We now have one equation that is linear in x; that is, the equation contains only the variable x.

Removing the parentheses and solving for x, we find that

$$x + 22 - 8x = 8$$
$$-7x = 8 - 22$$
$$-7x = -14$$
$$x = 2$$

To get the corresponding value of y, we substitute x = 2 in y = 11 - 4x. The result is

$$y = 11 - 4(2)$$
$$= 11 - 8$$
$$= 3$$

Thus, the solution for the two original equations is x = 2 and y = 3.

Practice problems. Solve the following systems by the substitution method:

1. $2x - 9y = 1$
   $x - 4y = 1$

2. $2x + y = 0$
   $2x - y = 1$

3. $5r + 2s = 23$
   $4r + s = 19$

4. $t - 4v = 1$
   $2t - 9v = 3$

Answers:

1. x = 5
   y = 1

2. x = 1/4
   y = -1/2

3. r = 5
   s = -1

4. t = -3
   v = -1

Literal Coefficients

Simultaneous equations with literal coefficients and literal constants may be solved for the value of the variables just as the other equations discussed in this chapter, with the exception that the solution will contain literal numbers. For example, find the solution of the system:

$$3x + 4y = a$$

$$4x + 3y = b$$

We proceed as with any other simultaneous linear equation. Using the addition method, we may proceed as follows: To eliminate the y term we multiply the first equation by 3 and the second equation by -4. The equations then become

$$9x + 12y = 3a$$

$$-16x - 12y = -4b$$

$$-7x \qquad = 3a - 4b$$

$$x \qquad = \frac{3a - 4b}{-7}$$

$$x = \frac{4b - 3a}{7}$$

365

To eliminate x, we multiply the first equation by 4 and the second equation by -3. The equations then become

$$12x + 16y = 4a$$
$$-12x - 9y = -3b$$
$$\overline{}$$
$$7y = 4a - 3b$$
$$y = \frac{4a - 3b}{7}$$

We may check in the same manner as that used for other equations, by substituting these values in the original equations.

INTERPRETING EQUATIONS

Recall that the general form for an equation in the first degree in one variable is $ax + b = 0$. The general form for first-degree equations in two variables is

$$ax + by + c = 0.$$

It is interesting and often useful to note what happens graphically when equations differ, in certain ways, from the general form. With this information, we know in advance certain facts concerning the equation in question.

LINES PARALLEL TO THE AXES

If in a linear equation the y term is missing, as in

366

$$2x - 15 = 0$$

the equation represents a line parallel to the Y axis and $7\frac{1}{2}$ units from it. Similarly, an equation such as

$$4y - 9 = 0$$

which has no x term, represents a line parallel to the X axis and $2\frac{1}{4}$ units from it. (See fig. 12-8.)

The fact that one of the two variables does not appear in an equation means that there are no limitations on the values the missing variable can assume. When a variable does not appear, it can assume any value from zero to plus or minus infinity. This can happen only if the line represented by the equation lies parallel to the axis of the missing variable.

Lines Passing Through the Origin

A linear equation, such as

$$4x + 3y = 0$$

that has no constant term, represents a line passing through the origin. This fact is obvious since x = 0, y = 0 satisfies any equation not having a constant term. (See fig. 12-8.)

Lines Parallel to Each Other

An equation such as

367

$$3x - 2y = 6$$

has all possible terms present. It represents a line that is not parallel to an axis and does not pass through the origin.

Equations that are exactly alike, except for the constant terms, represent parallel lines. As shown in figure 12-8, the lines represented by the equations

$$3x - 2y = -18 \text{ and } 3x - 2y = 6$$

are parallel.

Parallel lines have the same slope. Changing the constant term moves a line away from or toward the origin while its various positions remain parallel to one another. Notice in figure 12-8 that the line $3x - 2y = 6$ lies closer to the origin than $3x - 2y = -18$. This is revealed at sight for any pair of lines by comparing their constant terms. That one which has the constant term of greater absolute value will lie farther from the origin. In this case $3x - 2y = -18$ will be farther from the origin since $|-18| > |6|$.

The fact that lines are parallel is indicated by the result when we try to solve two equations such as $3x - 2y = -18$ and $3x - 2y = 6$ simultaneously. Subtraction eliminates both x and y immediately. If both variables disappear, we cannot find values for them such that both equations are satisfied at the same time. This means that there is no solution. No solution implies that there is no point of intersection for the straight lines represented by the equations. Lines that do not intersect in the finite plane are parallel.

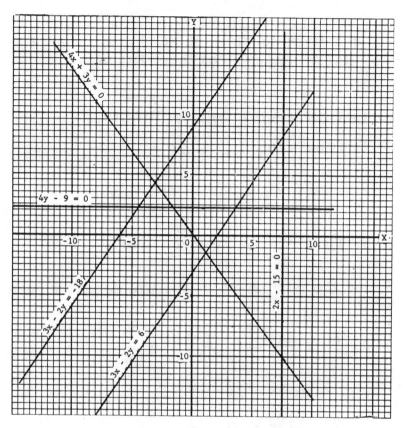

Figure 12-8.—Interpreting equations.

USING TWO VARIABLES IN SOLVING WORD PROBLEMS

Many problems can be solved quickly and easily using one equation with one variable. Other problems that might be rather difficult to solve in terms of one variable can easily be solved using two equations and two variables. The difference in the two methods is shown in the following example, solved first by using one variable and then using two.

369

EXAMPLE: Find the two numbers such that half the first equals a third of the second and twice their sum exceeds three times the second by 4.

SOLUTION USING ONE VARIABLE

1. Let x = the first number.

2. Then $\frac{x}{2} = \frac{1}{3}$ of the second number.

3. Thus $\frac{3x}{2}$ = the second number.

From the statement of the problem, we then have

$$2 \left(x + \frac{3x}{2} \right) = 3 \left(\frac{3x}{2} \right) + 4$$

$$2x + 3x = \frac{9x}{2} + 4$$

$$10x = 9x + 8$$

$$x = 8 \qquad \text{(first number)}$$

$$\frac{3x}{2} = 12 \qquad \text{(second number)}$$

SOLUTION USING TWO VARIABLES

If we let x and y be the first and second numbers, respectively, we can write two equations almost directly from the statement of the problem. Thus,

1. $\frac{x}{2} = \frac{y}{3}$

370

2. $2(x + y) = 3y + 4$

Solving for x in the first equation and substituting this value in the second, we have

$$x = \frac{2y}{3}$$

$$2 \left(\frac{2y}{3} + y \right) = 3y + 4$$

$$\frac{4y}{3} + 2y = 3y + 4$$

$$4y + 6y = 9y + 12$$

$$y = 12 \qquad \text{(second number)}$$

$$\frac{x}{2} = \frac{12}{3}$$

$$x = 8 \qquad \text{(first number)}$$

Thus, we see that the solution using two variables is more direct and simple. Often it would require a great deal of skill to manipulate a problem so that it might be solved using one variable; whereas the solution using two variables might be very simple. The use of two variables, of course, involves the fact that the student must be able to form two equations from the information given in the problem.

Practice problems. Solve the following problems using two variables:

1. A Navy tug averages 12 miles per hour downstream and 9 miles per hour upstream. How fast is the stream flowing?

2. The sum of the ages of two boys is 18. If 4

times the younger boy's age is subtracted from 3 times the older boy's age, the difference is 12. What are the ages of the two boys?

Answers:

1. $1\frac{1}{2}$ mph.

2. 6 years and 12 years.

INEQUALITIES IN TWO VARIABLES

Inequalities in two variables are of the following form:

$$x + y > 2$$

Many solutions of such an inequation are apparent immediately. For example, x could have the value 2 and y could have the value 3, since 2 + 3 is greater than 2.

The existence of a large number of solutions suggests that a graph of the inequation would contain many points. The graph of an inequation in two unknowns is, in fact, an entire area rather than just a line.

PLOTTING ON THE COORDINATE SYSTEM

It would be extremely laborious to plot enough points at random to define an entire area of the coordinate system. Therefore our method consists of plotting a boundary line and shading the area, on one side of this line, wherein the solution points lie.

The equation of the boundary line is formed by changing the inequation to an equation. For example, the equation of the boundary line for the graph of

$$x + y > 2$$

is the equation

$$x + y = 2$$

Figure 12-9 is a graph of $x + y > 2$. Notice that the boundary line $x + y = 2$ is not solid. This is intended to indicate that points on the boundary line are not members of the solution set. Every point lying above and to the right of the boundary line is a member of the solution set. Any solution point may be verified by substituting its X and Y coordinates for x and y in the original inequation.

SIMULTANEOUS EQUATIONS

The areas representing the solutions of two different inequations may overlap. If such an overlap occurs, the area of the overlap includes all points whose coordinates satisfy both inequations simultaneously. An example of this is shown in figure 12-10, in which the following two inequations are graphed:

$$x + y > 2$$
$$x - y > 2$$

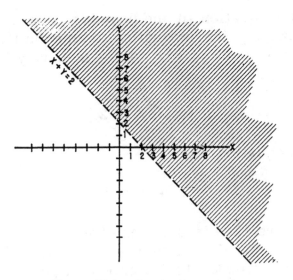

Figure 12-9.—Graph of x + y > 2.

The double crosshatched area in figure 12-10 contains all points which comprise the solution set for the system.

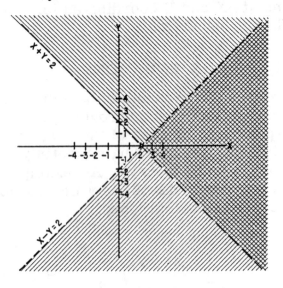

Figure 12-10.—Graph of x + y > 2 and x - y > 2.

CHAPTER 13

RATIO, PROPORTION, AND VARIATION

The solution of problems based on ratio, proportion, and variation involves no new principles. However, familiarity with these topics will often lead to quick and simple solutions to problems that would otherwise be more complicated.

RATIO

The results of observation or measurement often must be compared with some standard value in order to have any meaning. For example, to say that a man can read 400 words per minute has little meaning as it stands. However, when his rate is compared to the 250 words per minute of the average reader, one can see that he reads considerably faster than the average reader. How much faster? To find out, his rate is divided by the average rate, as follows:

$$\frac{400}{250} = \frac{8}{5}$$

Thus, for every 5 words read by the average reader, this man reads 8. Another way of making this comparison is to say that he reads $1\frac{3}{5}$

times as fast as the average reader.

When the relationship between two numbers is shown in this way, they are compared as a RATIO. A ratio is a comparison of two like quantities. It is the quotient obtained by dividing the first number of a comparison by the second.

Comparisons may be stated in more than one way. For example, if one gear has 40 teeth and another has 10, one way of stating the comparison would be 40 teeth to 10 teeth. This comparison could be shown as a ratio in four ways as follows:

1. 40:10

2. 40 ÷ 10

3. $\frac{40}{10}$

4. The ratio of 40 to 10.

When the emphasis is on "ratio," all of these expressions would be read, "the ratio of 40 to 10." The form 40 ÷ 10 may also be read "40 divided by 10." The form $\frac{40}{10}$ may also be read "40 over 10."

Comparison by means of a ratio is limited to quantities of the same kind. For example, in order to express the ratio between 6 ft and 3 yd, both quantities must be written in terms of the same unit. Thus the proper form of this ratio is 2 yd : 3 yd, not 6 ft : 3 yd. When the parts of the ratio are expressed in terms of the same unit, the units cancel each other and the ratio consists simply of two numbers. In this example, the final form of the ratio is 2 : 3.

Since a ratio is also a fraction, all the rules that govern fractions may be used in working with ratios. Thus, the terms may be reduced, increased, simplified, and so forth, according to the rules for fractions. To reduce the ratio 15:20 to lowest terms, write the ratio as a fraction and then proceed as for fractions. Thus, 15:20 becomes

$$\frac{15}{20} = \frac{3}{4}$$

Hence the ratio of 15 to 20 is the same as the ratio of 3 to 4.

Notice the distinction in thought between $\frac{3}{4}$ as a fraction and $\frac{3}{4}$ as a ratio. As a fraction we think of $\frac{3}{4}$ as the single quantity "three-fourths." As a ratio, we think of $\frac{3}{4}$ as a comparison between the two numbers, 3 and 4. For example, the lengths of two sides of a triangle are $1\frac{9}{16}$ ft and 2 ft. To compare these lengths by means of a ratio, divide one number by the other and reduce to lowest terms, as follows:

$$\frac{1\frac{9}{16}}{2} = \frac{\frac{25}{16}}{2} = \frac{25}{32}$$

The two sides of the triangle compare as 25 to 32.

INVERSE RATIO

It is often desirable to compare the numbers of a ratio in the inverse order. To do this, we simply interchange the numerator and the denominator. Thus, the inverse of 15:20 is 20:15. When the terms of a ratio are interchanged, the INVERSE RATIO results.

Practice problems. In problems 1 through 6, write the ratio as a fraction and reduce to lowest terms. In problems 7 through 10, write the inverse of the given ratio.

1. The ratio of 5 lb to 15 lb

2. $16 : $12

3. 16 ÷ 4

4. One quart to one gallon

5. 5x to 10x

6. $3\frac{1}{3} : 4\frac{1}{2}$

7. The ratio of 6 ft to 18 ft

8. $\frac{4}{8}$

9. 5 : 8

10. 15 to 21

Answers:

1. $\frac{1}{3}$ 2. $\frac{4}{3}$

3. $\frac{4}{1}$ 4. $\frac{1}{4}$

5. $\dfrac{1}{2}$ 6. $\dfrac{20}{27}$

7. $\dfrac{3}{1}$ 8. $\dfrac{2}{1}$

9. $\dfrac{8}{5}$ 10. $\dfrac{7}{5}$

PROPORTION

Closely allied with the study of ratio is the subject of proportion. A PROPORTION is nothing more than an equation in which the members are ratios. In other words when two ratios are set equal to each other, a proportion is formed. The proportion may be written in three different ways as in the following examples:

$$15:20 :: 3:4$$

$$15:20 = 3:4$$

$$\frac{15}{20} = \frac{3}{4}$$

The last two forms are the most common. All these forms are read, "15 is to 20 as 3 is to 4." In other words, 15 has the same ratio to 20 as 3 has to 4.

One reason for the extreme importance of proportions is that if any three of the terms are given, the fourth may be found by solving a simple equation. In science many chemical and physical relations are expressed as proportions. Consequently, a familiarity with proportions will provide one method for solving many applied problems. It is evident from the last form shown, $\frac{15}{20} = \frac{3}{4}$, that a proportion is really a fractional equation. Therefore, all the rules for fraction equations apply.

TERMS OF A PROPORTION

Certain names have been given to the terms of the two ratios that make up a proportion. In a proportion such as 3:8 = 9:24, the first and the last terms (the outside terms) are called the EXTREMES. In other words, the numerator of the first ratio and the denominator of the second are called the extremes. The second and third terms (the inside terms) are called the MEANS. The means are the denominator of the first ratio and the numerator of the second. In the example just given, the extremes are 3 and 24; the means are 8 and 9.

Four numbers, such as 5, 8, 15, and 24, form a proportion if the ratio of the first two in the order named equals the ratio of the second two. When these numbers are set up as ratios with the equality sign between them, the members will reduce to an identity if a true proportion exists. For example, consider the following proportion:

$$\frac{5}{8} = \frac{15}{24}$$

In this proportion, $\frac{15}{24}$ must reduce to $\frac{5}{8}$ for the proportion to be true. Removing the same factor from both members of $\frac{15}{24}$ we have

$$\frac{5}{8} = \frac{3(5)}{3(8)}$$

The number 3 is the common factor that must be removed from both the numerator and the denominator of one fraction in order to show that the expression

$$\frac{5}{8} = \frac{15}{24}$$

is a true proportion. To say this another way, it is the factor by which both terms of the ratio $\frac{5}{8}$ must be multiplied in order to show that this ratio is the same as $\frac{15}{24}$.

Practice problems. For each of the following proportions, write the means, the extremes, and the factor of proportionality.

1. $\frac{3}{16} = \frac{15}{80}$ 3. $\frac{25}{75} = \frac{1}{3}$

2. 4:5 = 12:15   4. 12:3 :: 4:1

   Answers:

1. Means: 16 and 15

   Extremes: 3 and 80

   Factor of proportionality: 5

2. M: 5 and 12

   E: 4 and 15

   FP: 3

3. M: 75 and 1

   E: 25 and 3

   FP: 25

4. M: 3 and 4

   E: 12 and 1

   FP: 3

OPERATIONS OF PROPORTIONS

It is often advantageous to change the form of a proportion. There are rules for changing or combining the terms of a proportion without altering the equality between the members. These rules are simplifications of fundamental rules for equations; they are not new, but are simply adaptations of laws or equations presented earlier in this course.

Rule 1. In any proportion, the product of the means equals the product of the extremes.

This is perhaps the most commonly used rule of proportions. It provides a simple way to rearrange a proportion so that no fractions are present. In algebraic language the rule is illustrated as follows:

$$\frac{a}{b} = \frac{c}{d}$$

$$bc = ad$$

To prove this rule, we note that the LCD of the two ratios $\frac{a}{b}$ and $\frac{c}{d}$ is bd. Multiplying both members of the equation in its original form by this LCD, we have

$$bd \cdot \frac{a}{b} = bd \cdot \frac{c}{d}$$

$$ad = bc$$

The following numerical example illustrates the simplicity of rule 1:

$$\frac{3}{8} = \frac{9}{24}$$

$$8(9) = 3(24)$$

If one of the terms of a proportion is a variable to the first power as in

$$7:5 = x:6$$

the proportion is really a linear equation in one variable. Such an equation can be solved for the unknown.

Equating the products of the means and extremes produces the following:

$$5x = 42$$

$$x = 8\frac{2}{5}$$

Mean Proportional

When the two means of a proportion are the same quantity, that quantity is called the MEAN PROPORTIONAL between the other two terms. In the proportion

$$\frac{a}{x} = \frac{x}{c}$$

x is the mean proportional between a and c.

Rule 2. The mean proportional between two quantities is the square root of their product. This rule is stated algebraically as follows:

$$\frac{a}{x} = \frac{x}{c}$$

$$x = \pm \sqrt{ac}$$

To prove rule 2, we restate the proportion and apply rule 1, as follows:

$$\frac{a}{x} = \frac{x}{c}$$

$$x^2 = ac$$

$$x = \pm \sqrt{ac}$$

Rule 2 is illustrated by the following numerical example:

$$\frac{2}{8} = \frac{8}{32}$$

$$8 = \sqrt{2(32)}$$

$$8 = \sqrt{64}$$

OTHER FORMS OF PROPORTIONS

If four numbers, for example, a, b, c, and d, form a proportion, such as

$$\frac{a}{b} = \frac{c}{d}$$

they also form a proportion according to other arrangements.

Inversion

The four selected numbers are in proportion by INVERSION in the form

$$\frac{b}{a} = \frac{d}{c}$$

The inversion relationship is proved as follows, by first multiplying both members of the original proportion by $\frac{bd}{ac}$:

$$\left(\frac{bd}{ac}\right)\left(\frac{a}{b}\right) = \left(\frac{bd}{ac}\right)\left(\frac{c}{d}\right)$$

$$\frac{d}{c} = \frac{b}{a}$$

Note that the product of the means and the product of the extremes still yield the same equality as in the original proportion.

The inversion relationship may be illustrated by the following numerical example:

$$\frac{5}{8} = \frac{10}{16}$$

Therefore,

$$\frac{8}{5} = \frac{16}{10}$$

Alternation

The four selected numbers (a, b, c, and d) are in proportion by ALTERNATION in the following form:

$$\frac{a}{c} = \frac{b}{d}$$

To prove the alternation relationship, first multiply both sides of the original proportion by $\frac{b}{c}$, as follows:

$$\frac{a}{b} = \frac{c}{d}$$

$$\frac{b}{c}\left(\frac{a}{b}\right) = \frac{b}{c}\left(\frac{c}{d}\right)$$

$$\frac{a}{c} = \frac{b}{d}$$

The following numerical example illustrates alternation:

$$\frac{5}{8} = \frac{10}{16}$$

Therefore,

$$\frac{5}{10} = \frac{8}{16}$$

SOLVING PROBLEMS BY MEANS OF PROPORTION

One of the most common types of problems based on proportions involves triangles with proportional sides. Suppose that the corresponding sides of two triangles are known to be proportional. (See fig. 13-1.) The lengths of the sides of one triangle are 8, 9, and 11. The length of the side of the second triangle corresponding to side 8 in the first triangle is 10. We wish to find the lengths of the remaining sides, b and c.

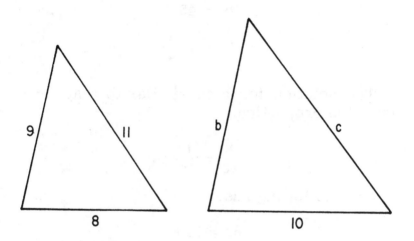

Figure 13-1.—Triangles with opposite sides proportional.

Since the corresponding sides are proportional, the pairs of corresponding sides may be used to form proportions as follows:

$$\frac{8}{10} = \frac{9}{b}$$

$$\frac{9}{b} = \frac{11}{c}$$

$$\frac{8}{10} = \frac{11}{c}$$

To solve for b, we use the proportion

$$\frac{8}{10} = \frac{9}{b}$$

and obtain the following result:

$$8b = 90$$
$$4b = 45$$
$$b = 11\frac{1}{4}$$

The solution for c is similar to that for b, using the proportion

$$\frac{8}{10} = \frac{11}{c}$$

with the following result:

$$8c = 110$$
$$c = 13\frac{3}{4}$$

The sides of the second triangle are 10, $11\frac{1}{4}$, and $13\frac{3}{4}$. The result can also be obtained by using the factor or proportionality. Since 8 and 10 are lengths of corresponding sides, we can write

$$8k = 10$$
$$k = \frac{10}{8} = \frac{5}{4}$$

The factor of proportionality is thus found to be $\frac{5}{4}$.

Multiplying any side of the first triangle by $\frac{5}{4}$ gives the corresponding side of the second triangle, as follows:

$$b = 9 \left(\frac{5}{4}\right) = \frac{45}{4} = 11\frac{1}{4}$$

$$c = 11 \left(\frac{5}{4}\right) = \frac{55}{4} = 13\frac{3}{4}$$

Proportional sides of similar triangles may be used to determine the height of an object by measuring its shadow. (See fig. 13-2.)

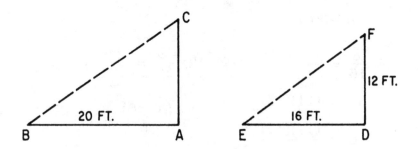

Figure 13-2.—Measuring height by shadow length.

In figure 13-2, mast AC casts a shadow 20 ft long when the shadow of DF is 16 ft long. Assuming that both masts are vertical and on level ground, triangle ABC is similar to triangle DEF and their corresponding sides are therefore proportional. Thus the height of AC may be found as follows:

$$\frac{AC}{12} = \frac{20}{16}$$

$$AC = \frac{(12)(20)}{16} = 15$$

Practice problems. In each of the following problems, set up a proportion and then solve for the unknown quantity:

1. Referring to figure 13-1, if the shortest side of the larger triangle is 16 units long, rather than 10, how long is side c?

2. If a mast 8 ft high casts a shadow 10 ft long, how high is a mast that casts a shadow 40 ft long?

Answers:

1. $\dfrac{8}{16} = \dfrac{11}{c}$

$8c = (11)(16)$

$c = \dfrac{(11)(16)}{8}$

$c = 22$

2. $\dfrac{8}{10} = \dfrac{h}{40}$

$\dfrac{(8)(40)}{10} = h$

$h = 32$

Word Problems

A knowledge of proportions often provides a quick method of solving word problems. The following problem is a typical example of the types that lend themselves to solution by means of proportion.

If an automobile runs 36 mi on 2 gal of gas,

how many miles will it run on 12 gal? Comparing miles to miles and gallons to gallons, we have

$$36:x = 2:12$$

Rewriting this in fraction form, the solution is as follows:

$$\frac{36}{x} = \frac{2}{12}$$

$$2x = 12(36)$$

$$x = 6(36)$$

$$= 216 \text{ mi}$$

Practice problems. In each of the following problems, first set up a proportion and then solve for the unknown quantity:

1. The ratio of the speed of one aircraft to that of another is 2 to 5. If the slower aircraft has a speed of 300 knots, what is the speed of the faster aircraft?

2. If 6 seamen can empty 2 cargo spaces in 1 day, how many spaces can 150 seamen empty in 1 day?

3. On a map having a scale of 1 in. to 50 mi, how many inches represent 540 mi?

Answers:

1. 750 kt   2. 50   3. 10.8 in.

VARIATION

When two quantities are interdependent, changes in the value of one may have a predictable effect on the value of the other. Variation is the name given to the study of the effects of changes among related quantities. The three types of variation which occur frequently in the study of scientific phenomena are DIRECT, INVERSE, and JOINT.

DIRECT VARIATION

An example of direct variation is found in the following statement: The perimeter (sum of the lengths of the sides) of a square increases if the length of a side increases. In everyday language, this statement might become: The longer the side, the bigger the square. In mathematical symbols, using p for perimeter and s for the length of the side, the relationship is stated as follows:

$$p = 4s$$

Since the number 4 is constant, any variations which occur are the results of changes in p and s. Any increase or decrease in the size of s results in a corresponding increase or decrease in the size of p. Thus p varies in the same way (increasing or decreasing) as s. This explains the terminology which is frequently used: p varies directly as s.

In general, if a quantity can be expressed in terms of a second quantity multiplied by a constant, it is said to VARY DIRECTLY AS the second quantity. For example if x and y are variables and k is a constant, x varies directly as y, if $x=ky$. Thus, as y increases x increases, and as y decreases, x decreases. There is a direct effect on x caused by any change in y.

The fact that x varies as y is sometimes indicated by $x \propto y$, or $x \sim y$. However, it is usually written in the form $x = ky$.

The relationship $x = ky$ is equivalent to $\frac{x}{y} = k$. If one quantity varies directly as a second quantity, the ratio of the first quantity to the second quantity is a constant. Thus, whatever the value of x, where it is divided by y, the result will always be the same value, k.

A quantity that varies directly as another quantity is also said to be DIRECTLY PROPORTIONAL to the second quantity. In $x = ky$, the coefficient of x is 1. The relationship $x = ky$ can be written in proportion form as

$$\frac{x}{k} = \frac{y}{1}$$

or

$$\frac{k}{x} = \frac{1}{y}$$

Notice that the variables, x and y, appear either in the numerators or in the denominators of the equal ratios. This implies that x and y are uirectly proportional. The constant, k, is the CONSTANT OF PROPORTIONALITY.

Practice problems. Write an equation showing the stated relationship, in each of the following problems:

1. The cost, C of a dozen wrenches varies directly as the price, p, of one wrench.

2. X is directly proportional to Y (use k as the constant of proportionality).

3. The circumference, C, of a circle varies directly as its diameter, d (use π as the constant of proportionality).

In the following problems, based on the formula p = 4s, find the appropriate word or symbol to fill the blank.

4. When s is doubled, p will be _____.

5. When s is halved, p will be _____.

6. _____ is directly proportional to s.

Answers:

1. C = 12p   4. doubled

2. X = kY   5. halved

3. C = πd   6. p

Variation as the Power of a Quantity

Another form of direct variation occurs when a quantity varies as some power of another. For example, consider the formula

$$A = \pi r^2$$

Table 13-1 shows the values of r and the corresponding values of A.

Table 13-1.—Relation between values of radius and area in a circle.

When r = ----	1	2	3	4	5	7	9
Then A = ----	π	4π	9π	16π	25π	49π	-81π

Notice how A changes as a result of a change in r. When r changes from 1 to 2, A changes from π to 4 times π or 2^2 times π. Likewise when r changes from 3 to 4, A changes not as r, but as the SQUARE of r. In general, one quantity varies as the power of another if it is equal to a constant times that quantity raised to the power. Thus, in an equation such as $x = ky^n$, x varies directly as the nth power of y. As y increases, x increases but more rapidly than y, and as y decreases, x decreases, but again more rapidly.

Practice problems.

1. In the formula $V = e^3$, how does V vary?

2. In the formula $A = s^2$, if s is doubled how much is A increased?

3. In the formula $s = \dfrac{gt^2}{2}$, g is a constant. If t is halved, what is the resulting change in s?

Answers:

1. Directly as the cube of e.

2. It is multiplied by 4.

3. It is multiplied by $\frac{1}{4}$.

INVERSE VARIATION

A quantity VARIES INVERSELY as another quantity if the product of the two quantities is a constant. For example, if x and y are variables and k is a constant, the fact that x varies inversely as y is expressed by

$$xy = k$$

or

$$x = \frac{k}{y}$$

If values are substituted for x and y, we see that as one increases, the other must decrease, and vice versa. Otherwise, their product will not equal the same constant each time.

If a quantity varies inversely as a second quantity, it is INVERSELY PROPORTIONAL to the second quantity. In xy = k, the coefficient of k is 1. The equality xy = k can be written in the form

$$\frac{x}{k} = \frac{1}{y}$$

or

$$\frac{k}{x} = \frac{y}{1}$$

Notice that when one of the variables, x or y, occurs in the numerator of a ratio, the other variable occurs in the denominator of the second ratio. This implies that x and y are inversely proportional.

Inverse variation may be illustrated by means of the formula for area of a rectangle. If A stands for area, L for length, and W for width, the expression for the area of a rectangle in terms of the length and width is

$$A = LW$$

Suppose that several rectangles, all having the same area but varying lengths and widths, are to be compared. Then LW = A has the same form as xy = k, where A and k are constants. Thus L is inversely proportional to W, and W is inversely proportional to L.

If the constant area is 12 sq ft, this relationship becomes

$$LW = 12$$

If the length is 4 ft, the width is found as follows:

$$W = \frac{12}{L} = \frac{12}{4} = 3 \text{ ft}$$

If the length increases to 6 ft, the width decreases as follows:

$$W = \frac{12}{6} = 2 \text{ ft}$$

397

If a constant area is 12, the width of a rectangle decreases from 3 to 2 as the length increases from 4 to 6. When two inversely proportional quantities vary, one decreases as the other increases.

Another example of inverse variation is found in the study of electricity. The current flowing in an electrical circuit at a constant potential varies inversely as the resistance of the circuit. Suppose that the current, I, is 10 amperes when the resistance, R, is 11 ohms and it is desired to find the current when the resistance is 5 ohms.

Since I and R vary inversely, the equation for the relationship is IR = k, where k is the constant voltage. Therefore, (10)(11) = k. Also, when the resistance changes to 5 ohms, (5)(I) = k. Quantities equal to the same quantity are equal to each other, so we have the following equation:

$$5I = (10)(11)$$

$$I = \frac{110}{5} = 22$$

The current is 22 amperes when the resistance is 5 ohms. As the resistance decreases from 11 to 5 ohms, the current increases from 10 to 22 amperes.

One type of variation problem which tends to be confusing to the beginner involves rates of speed or rates of doing work. For example, if 7 men can complete a job in 20 days, how long will 50 men require to complete the same job? The strictly mechanical approach to this problem might result in the following false solution, relating men to men and days to days:

$$\frac{7 \text{ men}}{50 \text{ men}} = \frac{20 \text{ days}}{T}$$

However, a little thought brings out the fact that we are dealing with an INVERSE relationship rather than a direct one. In other words, the more men we have, the less time is required. Therefore, the correct solution requires that we use an inverse proportion; that is, we must invert one of the ratios as follows:

$$\frac{7}{50} = \frac{T}{20}$$

$$T = \frac{(7)(20)}{50} = 2\frac{4}{5} \text{ days}$$

Practice problems. In problems 1 and 2, express the given data as a proportion, using k as the constant of proportionality.

1. The rate, r, at which a vessel travels in going a certain distance varies inversely as the time, t.

2. The volume, V, of a gas varies inversely as the pressure, p.

3. A ship moving at a rate of 15 knots requires 10 hr to travel a certain distance. If the speed is increased to 25 knots, how long will the ship require to travel the same distance?

Answers:

1. $\dfrac{r}{k} = \dfrac{1}{t}$

2. $\dfrac{V}{k} = \dfrac{1}{p}$

3. 6 hr

JOINT VARIATION

A quantity VARIES JOINTLY as two or more quantities, if it equals a constant times their product. For example, if x, y, and z are variables and k is a constant, x varies jointly as y and z, if x = kyz. Note that this is similar to direct variation, except that there are two variable factors and the constant with which to contend in the one number; whereas in direct variation, we had only one variable and the constant. The equality, x = kyz, is equivalent to

$$\frac{x}{yz} = k$$

If a quantity varies jointly as two or more other quantities, the ratio of the first quantity to the product of the other quantities is a constant.

The formula for the area of a rectangle is an example of joint variation. If A is allowed to vary, rather than being constant as in the example used earlier in this chapter, then A varies jointly as L and W. When the formula is written for general use, it is not commonly expressed as A = kLW, although this is a mathematically correct form. Since the constant of proportionality in this case is 1, there is no practical need for expressing it.

Using the formula A = LW, we make the following observations: If L = 5 and W = 3, then A = 3(5) = 15. If L = 5 and W = 4, then A = 4(5) = 20, and so on. Changes in the area of a rectangle depend on changes in either the length or the width or both. The area varies jointly as the length and the width.

As a general example of joint variation, consider the expression a ∝ bc. Written as an equation, this becomes a = kbc. If the value of a is known for particular values of b and c, we can find the new value of a corresponding to changes in the values of b and c. For example, suppose that a is 12 when b is 3 and c is 2. What is the value of a when b is 4 and c is 5? Rewriting the proportion,

$$\frac{a}{bc} = k$$

Thus

$$\frac{12}{(3)(2)} = k$$

Also,

$$\frac{a}{(4)(5)} = k$$

Since quantities equal to the same quantity are equal to each other, we can set up the following proportion:

$$\frac{a}{(4)(5)} = \frac{12}{(3)(2)}$$

$$a = 40$$

Practice problems. Using k as the constant of proportionality, write equations that express the following statements:

1. Z varies jointly as x and y.

2. S varies jointly as b times the square of r.

3. The length, W, of a radio wave varies jointly as the square root of the inductance, L, and the capacitance, C.

Answers:

1. $Z = kxy$

2. $S = kbr^2$

3. $W = k\sqrt{LC}$

COMBINED VARIATION

The different types of variation can be combined. This is frequently the case in applied problems. The equation

$$E = \frac{kw^2L}{p^2}$$

is an example of combined variation and is read, "E varies jointly as L and the square of W, and inversely as the square of p." Likewise,

$$V = \frac{krs}{t}$$

is read, "V varies jointly as r and s and inversely as t."

CHAPTER 14

DEPENDENCE, FUNCTIONS, AND FORMULAS

In chapter 13 of this course, use is made of several formulas, such as A = LW, E = IR, etc. It is the purpose of this chapter to explain the function and dependency relationships which make formulas so useful.

DEPENDENCE AND FUNCTIONS

Dependence may be defined as any relationship between two variables which allows the prediction of change in one of them as a result of change in the other. For example, the cost of 200 bolts depends upon the price per hundred. If C represents cost and p represents the price of 100 bolts, then the cost of 200 bolts may be expressed as follows:

$$C = 2p$$

In the example just given, C is called the DEPENDENT VARIABLE because its value depends upon the changing values of p. The INDEPENDENT VARIABLE is p. It is standard practice to isolate the dependent variable on the left side of an equation, as in the example.

Consider the formula for the area of a rectangle, $A = LW$. Here we have two independent variables, L and W.

Figure 14-1 (A) shows what happens if we double the length. Figure 14-1 (B) shows the result of doubling the width. Figure 14-1 (C) shows the effect of doubling both length and width. Notice that when the length or width alone is doubled the area is doubled, but when both length and width are doubled the area is four times as great.

In any equation showing a dependency relationship, the dependent variable is said to be a FUNCTION of the independent variable. Another use of the term "function" in describing an equation such as $C = 2p$ is to refer to the whole expression as "the function $C = 2p$." This terminology is especially useful when the right-hand expression has several terms. For example, consider the equation $y = 2x^2 + 3x - 4$. Mathematicians frequently use a shorthand notation and rewrite the equation as $y = f(x)$. The expression $f(x)$ is understood to mean "a function of x" and reference to the function by calling it $f(x)$ saves the space and time that would otherwise be required to write out all three terms.

Practice problems. Answer the following questions concerning the function $r = \dfrac{d}{t}$.

1. When t increases and d remains the same, does r increase, decrease, or remain the same?

2. When d increases and t remains the same, does r increase, decrease, or remain the same?

3. When t decreases and d remains the same, does r increase, decrease, or remain the same?

4. When d decreases and t remains the same, does r increase, decrease, or remain the same?

5. When d is doubled and t remains the same, is r doubled or halved?

6. When t is doubled and d remains the same, is r doubled or halved?

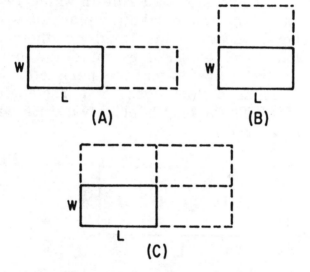

Figure 14-1.—Changes in the area of a rectangle resulting from changes in length and width.

Answers:

1. Decreases.   4. Decreases.

2. Increases.   5. Doubled.

3. Increases.   6. Halved.

405

FORMULAS

One of the most common uses of algebra is in the solution of formulas. Formulas have a wide and varied use throughout the Navy. It is important to know how formulas are derived, how to translate them into words, how to make them from word statements, and how to use them to solve problems.

A formula is a general fact, rule, or principle expressed in algebraic symbols. It is a shorthand expression of a rule in which letters and signs of operation take the place of words. The formula always indicates the mathematical operations involved. For example, the formula $P = 2L + 2W$ indicates that the perimeter (sum of the lengths of the sides) of a rectangle is equal to twice its length plus twice its width. (See fig. 14-2.)

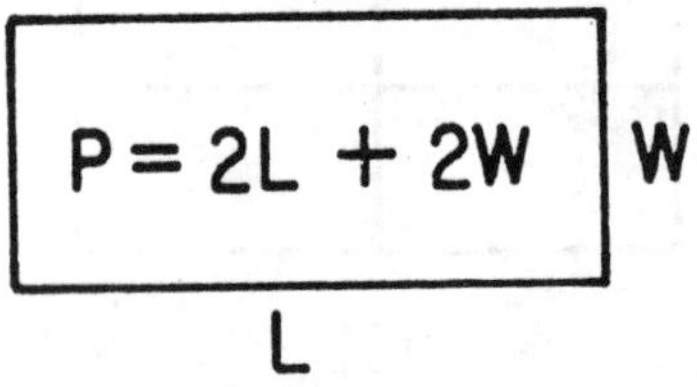

$$P = 2L + 2W \quad W$$

$$L$$

Figure 14-2.—Perimeter
of a rectangle.

A formula obtained by logical or mathematical reasoning is called a mathematical formula. A formula whose reliability is based on a limited number of observations, or on immediate experience, and not necessarily on established theories or laws is called an EMPIRI-

CAL formula. Empirical formulas are found frequently in engineering and physical sciences. They sometimes are valid for only a limited number of values.

SUBJECT OF A FORMULA

Usually a formula is taken almost directly from the verbal rule or law. For instance, the perimeter of a rectangle is equal to twice the length plus twice the width. Where possible, letters are used as symbols for the words. Thus, $P = 2L + 2W$. A simple formula such as this is like a declarative sentence. The left half is the SUBJECT and all the rest is the predicate. The subject is P. It corresponds to the part of the verbal rule that reads "the perimeter of a rectangle." This subject is usually a single letter followed by the equality sign.

All formulas are equations, but not all equations are formulas. Some distinctions between a formula and an ordinary equation are worthy of note. The equation may not have a subject, while the formula typically does. In the formula, the unknown quantity stands alone in the left-hand member. No computation is performed upon it, and it does not appear more than once. In the equation, on the other hand, the unknown quantity may appear once or more in either or both members, and computation may be performed with it or on it. We evaluate a formula by substituting for the literal numbers in the right member. An equation is solved by computation in either or both members until all that remains is an unknown in one member and a known quantity in the other. The solution of an equation usually requires a knowledge of algebraic principles, while the evaluation of a formula may ordinarily be accomplished with only a knowledge of arithmetic.

SYMBOLS

Letters that represent words have been standardized in many cases so that certain formulas may be written the same in various texts and reference books. However, to avoid any misunderstanding a short explanation often accompanies formulas as follows:

$$A = hw,$$

where

A = area in square units

h = height

w = width

Subscripts and Primes

In a formula in which two or more of the same kind of letters are being compared, it is desirable to make a distinction between them. In electronics, for example, a distinction between resistances may be indicated by R_a and R_b or R_1 and R_2. These small numbers or letters written to the right and below the R's are called subscripts. Those shown here are read: R sub a, R sub b, R sub one, and R sub two. Primes are also used in the same manner to distinguish between quantities of the same kind. Primes are written to the right and above the letters, as in S', S'', and S'''. They are read: S prime, S double prime, and S triple prime.

CHANGING THE SUBJECT OF A FORMULA

If values are given for all but one of its variables, a formula can be solved to obtain the value of that variable. The first step is usually the rearranging of the formula so that the unknown value is the subject—that is, a new formula is derived from the original. For example, the formula for linear motion—distance equals rate times time—is usually written

$$d = rt$$

Suppose that instead of the distance we wish to know the rate, r, or the time, t. We simply change the subject of the formula by the algebraic means developed in earlier chapters. Thus, in solving the formula for r, we divide both sides by t, with the following result:

$$\frac{d}{t} = \frac{rt}{t}$$

$$\frac{d}{t} = r, \text{ or } r = \frac{d}{t}$$

In words, this formula states that rate equals distance divided by time. Likewise, in solving for t, we have the following:

409

$$\frac{d}{r} = \frac{rt}{t}$$

$$\frac{d}{r} = t, \text{ or } t = \frac{d}{r}$$

In words, this formula states that time equals distance divided by rate.

We have in effect two new formulas, the subject of one being rate and the subject of the other being time. They are related to the original formula because they were derived from it, but they are different in that they have different subjects.

Practice problems. Derive new formulas from the following expressions with subjects as indicated:

1. $A = \frac{1}{2}bh$, subject h

2. $P = 2L + 2W$, subject L

3. $i = prt$, subject r

4. $p = br$, subject b

5. $E = IR$, subject I

6. The modern formula for converting Fahrenheit temperatures to Celsius (centigrade) is $C = (F + 40)\left(\frac{5}{9}\right) - 40$. Express the formula for converting Celsius (centigrade) temperatures to Fahrenheit.

Answers:

1. $h = \dfrac{2A}{b}$ 4. $b = \dfrac{p}{r}$

2. $L = \dfrac{P - 2W}{2}$ 5. $I = \dfrac{E}{R}$

3. $r = \dfrac{i}{pt}$ 6. $F = (C + 40)\dfrac{9}{5} - 40$

EVALUATING FORMULAS

The first step in finding the value of the unknown variable of a formula is usually the derivation of a formula that has the unknown as its subject. Once this is accomplished, the evaluation of a formula consists of nothing more than substituting numerical values for the letters representing known quantities and performing the indicated operations.

For example, suppose we wish to find the time required to fly 1,250 nautical miles at the rate of 250 knots. The formula is $d = rt$. We can change the subject by dividing both sides of the equation by r, as follows:

$$\frac{d}{r} = \frac{rt}{r}$$

$$\frac{d}{r} = t$$

$$t = \frac{1250}{250} = 5 \text{ hr}$$

Formulas can be solved for an unknown by substituting directly in the original formula even though that unknown is not the subject. Generally, however, it is simpler to first make the unknown the subject.

Formulas vary widely, from the simple type such as we have been considering to some that are very complex. All formulas have certain characteristics in common. There is always a subject, the quantity whose value is sought as a final answer. This subject usually stands alone, being placed equal to at least one and possibly several literal numbers, which are combined according to certain indicated operations. The formula can always be evaluated for a specific case when numerical values are known for all these literal quantities.

Evaluating formulas may be facilitated by developing a routine order of doing the work. If someone else can read the work and clearly understand what has been done, the work is in good order. The original formula should be written first, then the derived formula that is going to be used in solving the problem, and finally the actual substitutions. The indicated operations may then be carried out. Care should be taken to label answers with correct units; that is, miles per hour, foot-pounds, square feet, etc.

Practice problems.

1. $E = IR$. Solve for R in ohms if E is 110 volts and I is 5 amperes.

2. $d = rt$. Solve for t in hours if d is 840 nautical miles and r is 25 knots.

3. $F = (C + 40)\left(\dfrac{9}{5}\right) - 40$. Solve for C if F is $32°$.

Answers:

1. 22 ohms   2. 33.6 hr   3. $0°$

DEVELOPING FORMULAS

Developing a formula from a verbal statement is nothing more than reducing the statement to a shorthand form and showing the mathematical relationships between the elements of the statement.

For example, suppose that we wish to develop a formula showing the distance, D, traveled at the rate of 20 knots for t hours. If the distance traveled in 1 hr is 20 nautical miles, then the distance traveled in t hours is 20t. Therefore. the formula is

$$D = 20t$$

Practice problems.

1. Write a formula for the cost, C of p pounds of sugar at 15 cents per pound.

2. Write the formula for the cost, C, of one article when the total cost, T, of n similar articles is known.

3. Write a formula for the number of days, d, in w weeks.

4. Write a formula for the number of ounces, n, in p pounds.

Answers:

1. C = 15p   3. d = 7w

2. C = T/n   4. n = 16p

Developing Formulas from Tables

In technical work, instrument readings and other data are often recorded in a tabular arrangement. By careful observation of such tables of data, it is frequently possible to find values that are related in a definite pattern. The table can thus be used in developing a formula showing the relationship between the related quantities.

For example, table 14-1 shows the results of time trials on a ship, with the data rounded to the nearest whole hour and the nearest whole mile.

Table 14-1.—Time trials.

Nautical miles (d)	20	40	60	80	100
Hours (t)	1	2	3	4	5

By inspection of the table, it soon becomes clear that the number of miles traveled is always 20 times the corresponding number of hours. Therefore the formula developed from this table is as follows:

$$d = 20t$$

414

A second example of the derivation of a formula from a table is shown in figure 14-3. Figure 14-3 (A) shows several polygons (many-sided plane figures), each with one or more diagonals. A diagonal is a straight line joining one vertex (point where two sides meet) with another.

(A)

DIAGONALS (d)	1	2	3	4
SIDES (n)	4	5	6	7

(B)

Figure 14-3.—Diagonals of plane figures.

The table in figure 14-3 (B) compares the number of sides of each polygon with the number of diagonals that can be drawn from any one vertex. Using this table, we make a formula for the number, d, of diagonals that can be drawn from one vertex of a polygon of n sides. In the table we note that the number of diagonals is always 3 less than the number of sides. Therefore the formula is $d = n - 3$.

Practice problems. Complete the following

tables and write formulas to show the relationship between the numbers.

1.

L	2	5	8	11	14	17	20
P	12	30	48	66	84		

2.

a	0	1	2	3	4	5	6	7
b	4	5	6	7				

3.

x	0	1	2	3	4	5	6
y	0	3	6	9	12		

4.

n	1	2	3	4	5	6	7
s	3	6	11	18	27	38	

Answers:

1. $P = 6L$

3. $y = 3x$

2. $b = a + 4$

4. $s = n^2 + 2$

TRANSLATING FORMULAS

Thus far, we have been concerned primarily with reducing verbal rules or statements to formula form. It is also necessary to be able to do the reverse, and translate a formula into words. Technical publications frequently take advantage of the fact that it is more convenient to write formulas than longhand rules. Understanding is hampered if we are not able to translate these formulas into words. As an example of translation, we may translate the formula $V = lwh$ into words, with the literal factors representing words as follows:

V = volume of a
rectangular solid

l = length

w = width

h = height

This produces the following translation: The volume of a rectangular solid equals the length times the width times the height.

As a second example, we translate the algebraic expression $2\sqrt{x} - 4$ into words as follows: Twice the square root of a certain number, minus 4.

Practice problems. Translate each of the following expressions into words.

1. $PV = k$, where P represents pressure of a gas and V represents volume. (Assume constant temperature.)

2. $x = y + 4$, where x and y are numbers.

3. $A = LW$, where A is the area of a rectangle, L is its length, and W is its width.

4. $d = rt$, where d is distance, r is rate, and t is time.

Answers:

1. The pressure of a gas multiplied by its volume is constant, if the temperature is constant.

2. A certain number, x, is equal to the sum of another number, y, and 4.

3. The area of a rectangle is equal to the product of its length times its width.

4. Distance is equal to rate multiplied by time.

GRAPHING FORMULAS

We have seen that the formula is an equation. Since all formulas are equations they may be graphed. Graphs of formulas have wide use in the Navy in such fields as electronics and engineering. In practical applications it is often convenient to derive information from graphs of formulas rather than from formulas directly.

As an example, suppose that a fuel costs 30 cents per gallon. The formula for the cost in dollars of n gallons is

$$C = 0.30n$$

We see that this is a linear equation, the resulting curve of which passes through the origin (no constant term). Since we are interested only in positive values, we can eliminate three quadrants of the graph and use only the first quadrant. We already know one point on the graph is (0,0). We need plot only one other point to graph the formula. The result is shown in figure 14-4.

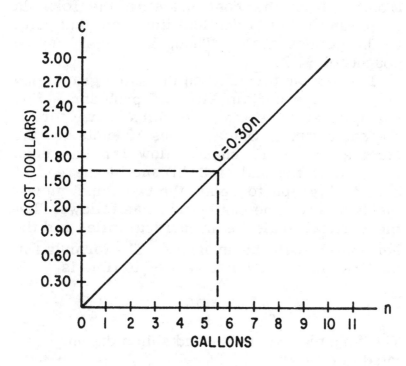

Figure 14-4.—Graph for the formula $C = 0.30n$.

We may read the cost directly from the graph when the number of gallons is known, or the number of gallons when the cost is known. For instance, if 5-1/2 gal are sold, find 5-1/2 on the gallons scale and follow the vertical line from that point to the point where it intersects the graph of the formula. From this point, follow the horizontal line to the cost scale. The horizontal line intersects the cost scale at 1.65. Therefore the cost of 5-1/2 gal is $1.65.

Likewise, to answer the question, "How many

419

gallons may be bought for $1.27," we would enlarge the graph enough to estimate to the exact cent. Then we would follow a horizontal line from 1.27 on the cost scale to the formula graph and follow a vertical line from that point to the gallons scale. Thus, 4-1/4 gal may be bought for $1.27.

Plotting two formulas on the same graph may help to solve certain kinds of problems. For example, suppose that two ships leave port at the same time. One averages 10 knots and the other averages 15 knots. How far has each traveled at the end of 3 hr and at the end of 5 hr? A graph to relate the two ships' movements at any time can be made as follows: Let the vertical scale be in nautical miles and the horizontal scale be in hours. The formula for the first ship's distance related to time is

$$d = 10t$$

The formula for the second ship's distance related to time is

$$d = 15t$$

We see that these formulas are linear and their curves pass through the origin. They are graphed in figure 14-5.

With this graph we can now answer the questions originally posed, at a glance. Thus in 3 hr the first ship traveled 30 mi and the second traveled 45 mi. In 5 hr the first ship traveled 50 mi and the second traveled 75 mi.

We could also answer such questions as:

420

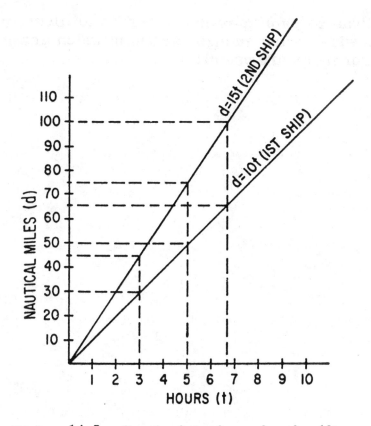

Figure 14-5.—Graph of the formulas d = 10t
and d = 15t.

When the second ship has traveled 100 mi, how
far has the other traveled? We first find the
point on the graph of d = 15t where the ship has
traveled 100 mi. We then follow the vertical
line from that point to the point where it intersects the graph of the other formula. From the
point of intersection we follow a horizontal line
to the distance axis and see that the first ship
has traveled about 67 mi when the second has
traveled 100 mi.

The foregoing examples serves to illustrate the wide variety of applications in which graphs of formulas are useful.

REA's **Problem Solvers**

The "PROBLEM SOLVERS" are comprehensive supplemental text-books designed to save time in finding solutions to problems. Each "PROBLEM SOLVER" is the first of its kind ever produced in its field. It is the product of a massive effort to illustrate almost any imaginable problem in exceptional depth, detail, and clarity. Each problem is worked out in detail with a step-by-step solution, and the problems are arranged in order of complexity from elementary to advanced. Each book is fully indexed for locating problems rapidly.

ACCOUNTING	LINEAR ALGEBRA
ADVANCED CALCULUS	MACHINE DESIGN
ALGEBRA & TRIGONOMETRY	MATHEMATICS for ENGINEERS
AUTOMATIC CONTROL	MECHANICS
SYSTEMS/ROBOTICS	NUMERICAL ANALYSIS
BIOLOGY	OPERATIONS RESEARCH
BUSINESS, ACCOUNTING, & FINANCE	OPTICS
CALCULUS	ORGANIC CHEMISTRY
CHEMISTRY	PHYSICAL CHEMISTRY
COMPLEX VARIABLES	PHYSICS
DIFFERENTIAL EQUATIONS	PRE-CALCULUS
ECONOMICS	PROBABILITY
ELECTRICAL MACHINES	PSYCHOLOGY
ELECTRIC CIRCUITS	STATISTICS
ELECTROMAGNETICS	STRENGTH OF MATERIALS &
ELECTRONIC COMMUNICATIONS	MECHANICS OF SOLIDS
ELECTRONICS	TECHNICAL DESIGN GRAPHICS
FINITE & DISCRETE MATH	THERMODYNAMICS
FLUID MECHANICS/DYNAMICS	TOPOLOGY
GENETICS	TRANSPORT PHENOMENA
GEOMETRY	VECTOR ANALYSIS
HEAT TRANSFER	

If you would like more information about any of these books,
complete the coupon below and return it to us or visit your local bookstore.

RESEARCH & EDUCATION ASSOCIATION
61 Ethel Road W. • Piscataway, New Jersey 08854
Phone: (732) 819-8880 **website: www.rea.com**

Please send me more information about your Problem Solver books

Name _____

Address _____

_____ State _____ Zip _____